AF522292

Challenges of Fisheries Sector

Challenges of Fisheries Sector

Mirza Akbar Khan

RANDOM PUBLICATIONS
NEW DELHI (INDIA)

Challenges of Fisheries Sector

ISBN 978-93-5111-147-4

Published in 2013 in India by
Reprinted, 2018

RANDOM PUBLICATIONS
4376-A/4B, Gali Murari Lal, Ansari Road
New Delhi-110 002
Phone : +9111-43580356, 011-23289044
e-mail: randompublications@hotmail.com

Type Setting by : Friends Media, Delhi-110089
Printed at : Replika Press Pvt. Ltd.

Preface

Technological advancements fuelled the growth of fisheries sector in India. Over eight fold increase in fish production during the last six decades was primarily due to transformation of traditional fishing practices into capital intensive industry through development and transfer of fisheries and aquaculture technologies. The available literature on the research and development process of Indian fisheries sector is scattered.

Fisheries sector can play a vital role, as a potential source, in attaining nutritional security in India. The current production of fish in India forms only 71.0% of the total potential and hence there is an ample scope of improvement and thus can be added up to the nutritional security. It is estimated that globally, about a third of the fish catch is not utilized for human food consumption because of post harvest loss resulting from poor handling and preservation. In India, there is also the problem of under utilization of by- catch due to the many species that are caught in the net. The cost effective and efficient utilization of aquatic products demands proper processing and distribution. Demand for fish and fish related products are increasing day by day in our country and reduction in post harvest losses can make a major contribution to satisfying this demand, improving quality and quantity for consumers and increasing income for the producers.

Integrated fish farming, an efficient method of combining fisheries, agriculture and livestock to increase the overall production and obtain maximum utilization of resources is a traditional practice in China. High fish production has been achieved. In order to transfer and adapt the Chinese technology in developing countries having different agro-climatic conditions, there is a need to establish through research a better understanding of the biological processes that are taking place in fish ponds under the system of integrated fish farming.

This book emphasized on fishery organizations, policies and various programmes made b central and state Govt. the inland and coastal aquaculture, socio economic issues of the fishery sector.

—Author

Contents

Preface ..... *v*

1. **Introduction to Fishes** **1**

Commercial and Subsistence Fishers 1
Status and Challenges of Fisheries 12

2. **Challenging Role of Fisheries Sector** **27**

Fisheries Sector 27
Role of Tifac in Fisheries Sector 28
Postmortem Changes in Fisheries Sector 32
Sensory Evaluation of Raw Fish in Markets 34
Fish Lipids of Importance 44
Amount of Vitamins and Minerals 54
Aquatic Resources and their Utilization 60
The Bacterial Flora on Live Fish 63

3. **Integrated Farming in Fisheries Sector** **71**

Integrated Fish Farming 71
Feeding Habits of Common Carp and Crucian Carp 77
Biology of Artificial Propagation 83
The Structure of the Testes and Sperm and the Development Stage of the Testes 87
Relationship between Endocrine System and Gonad Development 90
The Advantages of Integrated Fish Farming 94
Application of New Technology Stimulated and Improved the Production of Integrated Fish Farming 98
fish-cum-duck Integration of Fish Farming 114
The Economic Efficiency of Pig-grass-fish Integration 126
Appraisal of Economic Returns of Integrated Fish Farming 130

4. **Challenges of Fisheries in Biotechnology** **132**

Introduction 132
The Patterns in Developing Countries 134
Conservation and Utilization of Genetic Resources 139

Biotechnology in Fisheries Development 141
Biotechnology in Selected African Countries 143
Future Prospects and Strategies 150
Specific Spoilage Organisms 155

5. Techniques in Fish Handling and Processing 157

Handling Wet Fish 157
Freezing and Cold Storage of Fish 159
Handling Wet Fish on Shore 163
Handling Shellfish 165
The Importance of Containers in Fish Handling 167
Hygiene Requirements for Different Packing Materials 175
Fish Boxes in Developing Countries 180
The Effect of Hygiene during Handling 184
The Effect of Fish Species, Fishing Ground and Season 189

6. Genetic Biotechnologies in Fishery Sector 195

Management of Genetic Resources 195
Biotechnologies in Fishery Management 196
Biotechnology in Fisheries and Aquaculture 200
Aquaculture and Mrine Biotechnology 205
Potential and Limitations of Appropriate Biotechnologies 211
Biotechnology for the Enhancement of Feed Production 220
Recombinant DNA Technology and its Applications 222
Commercialization and Role of the Private Sector 225
Biotechnology Research and Development in the Private Sector 235

7. Fish Production Sector in Transitions Period 240

Proportion of Private Ownership 240
Probably Consists of Fish from a Mixture of Distinct Populations 243
Developing a Green-field Site 247
Finding and Identifying a Green-field Site Suitable for a Fishery 248
Recent Trends in Fisheries Production 257

***Index* 263**

1

Introduction to Fishes

A fish is any member of a paraphyletic group of organisms that consist of all gill-bearing aquatic craniate animals that lack limbs with digits. Included in this definition are the living hagfish, lampreys, and cartilaginous and bony fish, as well as various extinct related groups.

Most fish are ectothermic ("cold-blooded"), allowing their body temperatures to vary as ambient temperatures change, though some of the large active swimmers like white shark and tuna can hold a higher core temperature. Fish are abundant in most bodies of water. They can be found in nearly all aquatic environments, from high mountain streams (*e.g.*, char and gudgeon) to the abyssal and even hadal depths of the deepest oceans (*e.g.*, gulpers and anglerfish). At 32,000 species, fish exhibit greater species diversity than any other group of vertebrates. Fish are an important resource worldwide, especially as food.

COMMERCIAL AND SUBSISTENCE FISHERS

Commercial and subsistence fishers hunt fish in wild fisheries or farm them in ponds or in cages in the ocean. They are also caught by recreational fishers, kept as pets, raised by fishkeepers, and exhibited in public aquaria. Fish have had a role in culture through the ages, serving as deities, religious symbols, and as the subjects of art, books and movies. Fish are a huge group of animals, represented by over 28 000 different species. They are the oldest as well as the largest group of vertebrates. They are characterised by being almost exclusively aquatic throughout their lives. They are limbless, but have two sets of paired fins as well as a variety of single fins. They breathe predominantly by using organs known as gills. They are cold-blooded (poikilothermic), which means their body temperature varies with that of their surroundings. Their skin is usually covered with scales. Most temperate freshwater fish are called teleost (' true bone ') fish, which means that they have a bony skeleton. The only fish found in temperate freshwater that do not belong to this group are the sturgeons and the lampreys. Neither of these groups is found commonly in still-waters, although a few species of sturgeons have been introduced into still-waters in recent years. Teleost fish are extremely successful and well adapted to their environment.

FISH S PECIES

There are around 50 species of fish found for at least part of their lives in the freshwaters of the British Isles. This is quite a small number of species considering the diversity of the aquatic habitats available to them and the number of species in similar waters in continental Europe. Although many of the species are native to the British Isles, there are also many that have been successfully introduced over the centuries. Some of these have settled and become well established and are now generally considered ' native ', for example the carp. Some, however, still raise concerns among conservationists about the potential impact they may have, should they become more widespread; the zander and wels catfish for example, are described as ' non-native ' and are subject to regulation. Fish species have developed and evolved to make the best use of certain habitats. Many of the species present in the British Isles are especially developed to live in rivers. These species are usually unsuited to living in still-waters as they are not adapted to this very different type of habitat. In this chapter we will concentrate on those species already adapted to the still-water, and those that are commonly stocked in still-waters with apparent success, such as the chub. The barbel has been included, as it is stocked commonly in still-waters, but it should be borne in mind that special conditions must be maintained for this species owing to its requirement for clear, oxygen-rich water. This chapter sets out some detailed information about each species, which should allow fishery managers to understand better the species under their custody. The ecology of lakes, should allow the manager to establish the best conditions for the fish species and assess any other species for likely success in any given situation.

The fish described in this chapter are the following:

Family Cyprinidae

- Barbel
- Bream
- Silver bream
- Common carp
- Chub
- Crucian carp
- Goldfish.

Cyprinidae

The barbel is a large river fish with a long, streamlined, slightly flattened body. The juveniles are an overall brown colour but as they get larger they develop a dark brown back with golden bronze flanks and a creamy white underbelly. Adults generally grow to around 70 cm, but can get up to 90 cm and weigh in excess of 6 kg (the British rod-caught record 2008 is 9.837 kg). The barbel is native to the southern UK and mainland Europe. It is widespread and relatively common.

The barbel will generally mature after 3-5 years. Spawning takes place once a year in the late spring, usually May-July. Barbel spawn in small groups among gravel and stones in fast-flowing rivers. The adhesive eggs stay within the gravel bed until they hatch 10-15 days later. The fry tend to congregate in slow flowing areas where they feed on zooplankton. As they mature and get larger they tend to move out into the main flow of the river, feeding on invertebrates and even small fish in among clean gravel and associated vegetation.

The barbel is a very popular coarse fish, being one of the most sought-after river fish. They are a hard fighting fish that can grow very large, attracting specialist anglers.

Barbel have increasingly been stocked in still-waters over several years. They appear to grow and survive reasonably well, but there is little evidence that they can successfully spawn in most still-waters. It is vital that any stocking is with barbel obtained from a reputable fish farm. Barbel can sometimes be introduced from fish poached from the wild. Not only is this very harmful to the wild populations but evidence suggests that these fish find it very difficult to make the transition from rivers to still-waters. Generally barbel should not be stocked in shallow still-waters with a predominance of bottom feeders such as carp.

Perch (Perca fluviatilis)

The perch is a beautiful, deep-bodied fish with four to six dark bands over olive-green flanks. The abdomen is creamy silver with prominent red pelvic and anal fins. It can grow to 20 – 40 cm and weigh 1 – 2.5 kg (the British rod-caught record 2008 is 2.523 kg). The perch is native to the UK and Europe and is now widespread and common in the UK, northern Europe and Asia. It is a popular food fish in many countries; in England it used to be canned and sold as ' perchines ' .

Perch generally mature after 3 – 4 years and live to around 10 years. They spawn in early spring, usually March to May. Perch tend to spawn in the margins, producing long strings of eggs which lie wrapped around submerged vegetation. The eggs hatch after 10 days and the juveniles feed on small invertebrates as well as fish fry, including other perch fry. As they get older they start to feed on larger invertebrates and begin to feed actively on other fish. Small perch shoal in groups but as they get bigger they do become more solitary, with large individuals tending to move around in small shoals of just a few individuals.

The perch is a popular fish for coarse fisherman, being one of the most attractive of the freshwater fish species. It responds well to anglers ' baits and is readily caught. As a predator it is a useful fish to control the populations of other fish species. However, they can mature at a very small size (10 – 15cm) and in certain circumstances they can create an abundance of small perch. It appears that perch only grow to a large size where there is a plentiful supply of small prey fish.

Zander (Sander lucioperca)

The zander is a large predatory species with a long, thick body. The back is a dark brown – green with obvious dark brown vertical stripes running to the middle of the body. It grows quite large, reaching average sizes of 50 – 70cm, occasionally much bigger to a maximum of 120 cm (the British rod-caught record 2008 is 8.7 kg). The zander is common throughout Europe. It was deliberately introduced into England (the River Great Ouse) in 1963 and has spread widely since.

Zander generally mature after 3 – 5 years. Spawning takes place from April to June in among marginal vegetation. Pairs of fish spawn together at dawn and lay sticky eggs on vegetation and stones on the bottom. Unusually, both parents will guard the eggs until they hatch 5 – 10 days after spawning. The fry start on a diet of invertebrates but move quickly on to eat mainly of other fish. The zander is a hunter, patrolling open waters and chasing down small prey fish. Occasionally it appears to hunt as a group, although it mainly hunts as an individual.

The zander is a very popular fish with anglers, being a fast-growing, large and aggressive fish. It survives well in larger, clear still-waters and may be a useful addition to a specimen water as it prefers small fish as prey. Licences are required to stock this species.

Gasterosteidae

Three-s pined s tickleback *(Gasterosteus aculeatus)*

The stickleback is one of the most common and recognisable species in freshwaters. It is a small, thin-bodied fish with an olive-brown back and silvery sides and underside. The body is covered in large bony plates and there are three obvious strong spines on the back. The stickleback is a very small fish that rarely exceeds 10 cm in length (the British rod -caught record 2008 is 7 g (0.007 kg)). Sticklebacks are widely distributed and common throughout Europe. Sticklebacks mature after 1 – 2 years. Spawning takes place in March – June. They are unusual in that the male builds a nest of fibrous plant material and attracts females to spawn within it. The male also develops a vivid red throat to help attract a mate. The eggs take 10 – 20 days to hatch and are guarded constantly by the male. This guarding behaviour continues until the fry are a few weeks old. The fry feed on small invertebrates. As they get older they continue to feed on invertebrates but may extend their diet to include fish fry. Sticklebacks tend to aggregate into small shoals, only becoming solitary and territorial during spawning times.

The stickleback is not a target species for anglers. It is common in most still-waters and will almost certainly appear without help.

Nine-s pined s tickleback *(Pungitius p ungitius)*

This is a rare species. It is similar in most respects to the three-spined stickleback, except that it has 7 – 12 spines running along its back. It also goes dark with a black throat when in breeding condition.

Common Bream (Abramis brama)

The common bream is a large, bottom-feeding species with a deep thin body. When young it is a very silvery fish, but as it matures it becomes darker and more golden brown to green. It can grow to nearly a metre in length and weigh up to 8 kg (the British rod-caught record 2008 is 8.329 kg). The common bream is native to the UK and central Europe where it is widespread and common. It is a popular food fish in eastern Europe.

Bream generally mature after 4 years. Spawning takes place in May and June in dense vegetation in shallow water. The males develop white bumps (tubercles) over the body surface and become territorial. The females spawn at night in large groups and their sticky eggs attach to the vegetation. After 5 – 10 days the eggs hatch.

The fry start by feeding on zooplankton such as rotifers and then larger zooplankton such as daphnia. They quickly move on to their normal adult diet of benthic invertebrates, preferentially worms, molluscs and insect larvae such as chironomid midge larvae. The adults feed by using their extendable tubular mouth to root around in the pond bottom, taking large mouthfuls and ' winnowing ' out the food items. They are strongly shoaling fish from larval fry to late adulthood. Some of the very large individuals appear to be solitary or in small shoals. The common bream is one of the most popular of the coarse fish. It responds well to anglers ' baits and is readily caught. It grows quickly to a good size and survives well in many different conditions. As a bottom feeder it will disturb the bottom of the pond, causing suspended solids to cloud the water column. They are good in a mixed silver-fish population. The common bream should not be mixed with large numbers of carp as the species compete for the same food source.

The silver bream is a small mid-water species, with a deep, compressed body. When young it is a very silvery fish but as it matures its back becomes a darker olive/grey, although it still retains a very silvery look. Its maximum size is around 20 – 25 cm in length (the British rod-caught record 2008 is 0.425 kg). The silver bream is native to England, central and northern Europe, where it widespread and reasonably common. In England it is generally restricted to the southeast but is becoming more widespread.

Silver bream mature at around 3 – 4 years old. Spawning takes place in May – July in dense vegetation in shallow water. They spawn in groups early in the morning, scattering their sticky eggs among the submerged weeds. The eggs hatch in 4 – 8 days and begin feeding on small planktonic invertebrates, moving to larger invertebrates and plant material as they grow. Silver bream form strong shoals with individuals of a similar size and age throughout their lives. The silver bream is not a popular species for anglers, largely because it is commonly mistaken for a small common bream. Indeed it is very similar to a common bream of the same size, except the silver bream is slightly thicker in the body and has larger scales. They do make useful additions to a lake as they feed in a very different manner.

The carp is a large, bottom-feeding species with a thickset rounded body. They are usually a dark olive brown on the back, lightening to a light yellowish brown on the belly. It can grow to very large sizes, sometimes exceeding 1 m in length (the British rod-caught record 2008 is 26.9 kg). The carp is native to eastern Europe and Asia, but has now achieved worldwide distribution. It is very widespread and common in the UK, where it was first introduced in the nineteenth century.

Carp mature after 2 – 3 years. Spawning takes place in June to July in dense vegetation on the margins of the water body. The males develop small white lumps (tubercles) on the flanks and head. The females spawn at dawn and the sticky eggs attach to the weed. The eggs will hatch after 5 – 7 days and the fry start feeding on zooplankton, particularly rotifers. They quickly move to feeding on the bottom-dwelling invertebrates that make up the bulk of their diet. Their mouth is well designed for this feeding technique, being protrusible. However, carp are a very adaptable species and they can often be seen feeding on invertebrates on marginal vegetation and even off the surface. They tend to move around in small shoals of similar-sized individuals.

The carp is probably the most popular species for anglers on still-waters. It grows quickly to very large sizes and responds well to a variety of anglers ' baits. It is a very adaptable species so can survive and thrive in many different conditions. As a bottom feeder, it will disturb the sediment on the bottom of the water body, causing suspended solids to cloud the water. Carp have been bred for food over many centuries and it is one of the main global species for aquaculture in ponds. This has led to several different varieties of carp being commonly found. The most obvious of these varieties is the common carp, which is the fully scaled type. Mirror carp were bred to make it easier to remove the scales before cooking. Ornamental varieties are also common, from full-colour varieties, known as koi, to slightly gold-tinged cross breeds known as ghost carp. These are all the same species of fish, however.

The chub is a large, mid-water and surface-feeding fish with a rounded, streamlined body. When young it is a very silvery fish with an olive green back; as it gets larger its upper body darkens to a steel grey colour but retains the silvery flanks and belly. It is a large fish, attaining sizes of between 40 and 50 cm, rarely getting to up to 80 cm (the British rod-caught record 2008 is 3.9 kg). The chub is native to the UK and central Europe and is widespread and common. It is caught for food in some parts of Europe. Chub generally mature after 3 – 4 years. Spawning takes place during April – June among clean gravel and stones in flowing water. They spawn in the early morning in small groups of one female with several males in attendance. The eggs hatch within 6 – 8 days and start to feed on benthic invertebrates. As they grow older their diet changes to include larger invertebrates, fruit, vegetation and even small fish. When young, chub move around in shoals but as they mature they become much more solitary and even appear to be somewhat territorial. In rivers they migrate significant distances to feeding and spawning areas. Little is known about their activities in still-waters.

The chub is a popular fish among anglers because it is a large, hard-fighting fish that responds well to anglers ' baits. It grows quickly to a good size and appears to do very well in still-waters. Current evidence suggests that it does not successfully spawn in most still -waters, although it does spawn in large, clear-water lakes.

The crucian carp is a medium-sized, deep-bodied fish with marked lateral compression. Young fish are a golden bronze colour, which turns darker as they get older, with mature specimens being olive brown on the back through deep bronze on the flanks to a golden yellow on the belly. Adults can attain 20 – 30 cm in length and 1.5 kg in weight, although many crucian carp populations remain ' stunted ' with the mature adults rarely getting more than 15 – 20 cm and 0.5 kg (the British rod-caught record 2008 is 2.01 kg). The crucian carp is native to the UK, Europe and central Asia. There is some argument about whether it is a native to the UK and western Europe, but there seems little evidence to suggest it is introduced into these areas, where it is widespread. Crucian carp generally mature after 2 – 3 years. Spawning takes place in May – June in shallow water, in thick vegetation. The fish shed sticky eggs onto the vegetation where they take between 5 and 7 days to hatch, depending on the vegetation. Once hatched, and after the yolk sac has been used, the fry tend to form loose shoals and feed on small zooplankton. As they grow they start to feed mainly on invertebrates, such as molluscs and worms on the bottom of the pond, as well as plant material. Both the juveniles and the adults tend to live in loose shoals. They tend to occupy the mid-water and bottom of still-waters. Crucian carp prefer still-waters and particularly thrive in rich, lowland still-waters with abundant vegetation. Crucian carp have a remarkable ability to survive very low dissolved oxygen situations for long periods, leading to situations where they are often the only fish species to survive. Here they form healthy populations dominated by small ' stunted ' individuals. In larger waters they can grow much larger and seem to adopt a slightly different body shape.

This is a very popular angling species and is readily caught. It survives well in well -managed fisheries dominated by silver fish, but it is generally unsuitable for stocking in waters dominated by common carp. Crucians readily hybridise with goldfish and common carp. They are then displaced by the vigorous hybrid. They are also threatened by the non-native parasitic tapeworm *Bothriocephalus acheilognathi.*

The goldfish is a medium – sized mid-water species with a relatively deep and laterally compressed body. Wild goldfish are a deep brown colour throughout their lives, although there are many colour variants from deep orange through to almost white. Adults can get to around 30 cm and 1 kg in size. The goldfish is native to Asia, but owing to its ornamental value it has achieved worldwide distribution, being found on most continents. It is an introduced species in the UK.

Goldfish generally mature after 2 – 3 years. Spawning takes place during June and July in thick weed in the shallow margins. The eggs hatch in 4 – 7 days, depending on the water temperature, and the young feed on zooplankton. As they grow, the diet changes to small invertebrates and plant material. Goldfish tend to form loose shoals and prefer shallow, rich ponds with abundant vegetation. They are a hardy species, however, and appear to thrive in most still-water conditions.

The goldfish has become increasingly common in still-waters in the UK and Europe. It is apparently popular with some fishermen.

Goldfish readily hybridise with common carp and crucian carp, producing vigorous offspring that can successfully outcompete the original parent stock. Stocking with goldfish or hybrid strains is advised against.

The grass carp is a large mid-water feeding species with a solid, rounded body. It has a dark blue-black back, shading to silver on the belly. It can grow quite large, with adults reaching up to 100 cm in length (the British rod-caught record 2008 is 20.185 kg). Grass carp are originally from Asia, but have been widely distributed around the world as a method of weed control. They were introduced into England in the 1960s to control weed growth in some still-waters. Since then they have become widely distributed, but because of their specialised breeding requirements there appear to be no breeding populations.

Grass carp mature after 3 – 4 years. Spawning takes place in fast-flowing water in the headwaters of large rivers. Temperatures must be around 23 – 25 ° C. The eggs are unusual in that they are pelagic, being carried down river in mid-water until they hatch in the slower flows downstream. The fry start by feeding on small invertebrates, but as they get older they begin to feed on aquatic plants. Grass carp form loose shoals throughout their lives.

Grass carp are not particularly popular among anglers as they are generally difficult to catch and do not fight as hard as their size might suggest. They are not very effective in controlling weed growth as they do not feed until the waters are quite warm.

The gudgeon is a small, rounded, bottom-dwelling fish. Its back is a dark olive green to brown with lighter flanks and a silvery belly. It has dark markings along the flank and upper body. It is a very small fish, rarely reaching over 15 cm although it can grow to about 20 cm. (the British rod-caught record 2008 is 0.141 kg). Gudgeon are widely distributed throughout Europe and Asia and they are native to the UK.

Gudgeon mature at between 2 and 3 years old. They will spawn on stony or gravely bottoms, or among weed, and prefer flowing water, although they will spawn successfully in still-waters. They spawn during May or June and the eggs hatch in around 15 days. They feed on insects, molluscs, crustaceans and occasionally plant material. They are generally a solitary fish, but sometimes form into small shoals of perhaps a dozen individuals.

The gudgeon is not a popular target species for anglers, but it does do well in some still -waters where it probably makes an important contribution to the health of the lake.

The minnow is a small mid-water species with a long rounded body. It has a brown-gold body with numerous brown and black spots down the flanks usually joining to form an obvious black line along the body. It is a small fish, rarely exceeding 10 cm in length (the British rod-caught record 2008 is 0.135 kg). The minnow is a native to the UK, Europe and Asia where it is very widespread and common.

Minnows mature after 1 – 2 years. Spawning takes place in June or July, usually over clean stones or gravels. The males particularly become more colourful during spawning, with a scarlet belly and black throat. The eggs take 5 – 10 days to hatch and immediately start feeding on small invertebrates. As they grow they continue to feed on a variety of invertebrates. Minnows generally aggregate into small, loose shoals.

The minnow is not a target fish for anglers, although it can be important as a bait fish. It does well in lakes and ponds and can be a very useful addition to the lake ecology, particularly if predatory fish are present.

The orfe is a large, mid-water and surface-feeding fish with a rounded, streamlined body. When young it is a very silvery fish, but as it gets larger its upper body darkens to a brownish grey colour. Coloured ornamental varieties are common, particularly golden and blue orfe. It is a medium-sized fish, attaining sizes of between 30 and 40 cm, although it can occasionally get up to 80 cm. (the British rod-caught record 2008 is 3.7 kg). The orfe is native to central and eastern Europe and parts of Asia, where it is widespread and common. It is caught for food in some parts of Russia. It was introduced into England and is now widespread and common.

Orfe generally mature after 3 – 4 years. Spawning takes place during April and May, among weeds and clean stones in shallow water. The eggs hatch within 15 – 20 days and start to feed on benthic invertebrates. As they grow older their diet changes to include larger invertebrates, although they will take fruit, vegetation and even small fish. Orfe move around in shoals of similar-sized and aged individuals. They are found equally in rivers and still-waters.

The orfe is a popular fish among anglers because it is a good-sized, hard fighting fish that responds well to anglers ' baits. It grows quickly, to a good size and it appears to do very well in still-waters. The roach is medium-sized, mid-water fish with a relatively deep body. The young fish are silvery, but they quickly develop a dark blue – green back and light underside. Adults generally get to 20 – 40 cm and between 1 and 2 kg (the British rod-caught record 2008 is 1.899 kg). The roach is native to the UK and Europe, where it is very widespread and common throughout the lowland catchments.

Roach generally mature after 2 or 3 years. Spawning takes place once a year in the middle of spring, usually April – June. The roach spawn at dawn, shedding their sticky eggs on dense, submerged vegetation in the shallow areas of the lake.

The eggs will hatch in 5 – 10 days. Once the yolk sac is used, the fry will feed on zooplankton and other small invertebrates. The juveniles prefer to live in large shoals in the shallow, marginal areas with plenty of vegetation, which provide cover and food. As they mature and get larger they move out into open water, shoaling in smaller groups of similar-sized individuals. These larger individuals feed on larger benthic invertebrates and pondweed, particularly attached algae (periphyton).

They tend to occupy the mid-water to bottom of the pond. The roach is one of the most popular coarse fish in the UK. It responds well to anglers ' bait and is readily caught. It grows steadily and survives well in many different conditions. It is an excellent species in any mixed fishery as it is very adaptable. In the right conditions, it can quickly overpopulate with a very large population of small fish.

The rudd is a medium-sized, top-feeding species, with a deep, laterally compressed body. It has a dark greenish-brown back, grading through to a silvery underside. The flanks are golden brown. The iris is a deep gold with a red fleck on the upper side. The fins are normally a bright red. It can grow to 20 – 25 cm and 0.5 – 1 kg (the British rod-caught record 2008 is 2.1 kg). The rudd is native to the UK and central Europe, where it is very widespread and common.

Rudd generally mature at 2 – 4 years and live to 10 – 14 years. They spawn in late spring, usually from April to July. Rudd spawn by shedding their eggs on submerged vegetation around the margins. The eggs hatch in around 5 – 10 days and the fry stay within the protection of the vegetation, feeding on small zooplankton. As they grow they form compact shoals, feeding on the surface and in midwater. They prefer zooplankton, weed-dwelling invertebrates and terrestrial insects on the surface. Rudd appear always to spend their time in fairly compact shoals, no matter how large they get.

The rudd is a popular fish to catch, being attractive and readily taking bait, which is particularly important if other species are more reluctant to feed on anglers ' bait. It grows steadily but rarely attains large sizes. It is a very good fish for a mixed fishery, as it does not compete directly with any other fish species.

The tench is a large, bottom-feeding species, with a thickset, slightly laterally compressed body. The tench is a beautiful olive to dark green throughout its life, with small orange to red eyes. Adults generally grow slowly to a maximum length of 60 – 70 cm and can weigh up to 7 kg. (the British rod-caught record 2008 is 6.9 kg). The tench is native to the UK and Europe, although absent from the northern parts of both. The species has been widely introduced into Asia and North America.

Tench generally mature at 3 – 4 years old. Owing to their slow growth, they may only be 15 cm at this age. Spawning takes place once a year in spring, usually between May and July. They spawn among aquatic plants in shallow waters, to which their sticky eggs attach. The eggs will hatch in 4 – 5 days, and once the yolk sac is used the fry will feed on zooplank-ton, before switching to grazing among the bottom of water plants on a variety of aquatic invertebrates and attached algae. Tench prefer still-waters, or very slow flowing rivers with an abundance of macrophytes. The juveniles may form loose shoals, but as they mature they become more solitary, almost appearing to have preferred territories.

The tench is one of the most popular and easily recognisable of the coarse fish in the UK and Europe. Its popularity relates to its size and its beauty. It is generally a difficult fish to catch on rod and line. It is quite adaptable to a mixed fishery, but does not thrive in waters without good communities of aquatic macrophytes.

The pike is a large predatory fish, with a slim streamlined body and a large head with an enormous mouth full of teeth. It has a dark green, mottled back, grading through light green flanks to a creamy yellow underbelly. It is a very large fish, routinely growing to 50 – 100 cm and occasionally 140 cm and 30 kg (the British rod-caught record 2008 is 21.2 kg). The pike is widespread and abundant throughout the Northern Hemisphere in Europe, Asia and North America. It is a popular food fish wherever it is found.

Pike generally mature after 3 – 4 years, the males earlier and smaller than the females. They spawn among weed during February – May. The eggs attach to the weed and hatch after 10 – 15 days. When they hatch they start by feeding on invertebrates but quickly turn to small fish and then onto larger prey, including other vertebrates such as ducks and water rats. They are very partial to other pike and are, unsurprisingly, very solitary and territorial. It appears that females grow much larger than males, and it is said that any pike over 5 kg will be a female. They are an ambush predator, laying in wait for their prey and only striking at the very last minute when the prey is within one to two body lengths. The pike is one of the most popular freshwater fish in the UK. It grows exceptionally quickly to a large size and is a ferocious fighter, particular in the first moments as it attacks the bait.

Percidae

The ruffe is a small fish with an obvious spiny dorsal fin, large eyes and mouth. Its back is a dark green-brown colour with irregular dark blotches, whereas its underside is pale yellow to cream in colour. It usually grows to around 15 cm but can exceptionally grow to 25 cm (the British rod-caught record 2008 is 0.142 kg). The ruffe is native to much of Europe but was confined to eastern England, although its range has spread considerably in recent decades.

The ruffe matures after 1 to 2 years. They spawn in shallow water among stones and vegetation, mainly in April to May, although they probably spawn several times in a year. Initially the fry grow quickly on a diet of small invertebrates before moving on to larger invertebrates and other food such as fish eggs and young fry. The ruffe is a shoaling fish, moving around in groups throughout its life. The ruffe is not a popular fish for anglers, being such a small size and attacking baits destined for other larger target species. It has numerous alternative common names such as pope, tommy and daddie.

STATUS AND CHALLENGES OF FISHERIES

The fisheries administrators have viewed MCS as little more than the policing of the various maritime zones controlled by the State. This chapter attempts to provide a broader view of MCS - a view of MCS as the vital executive arm of fisheries management. The rapid depletion of key fish stocks in the 1980s and 1990s has made it imperative that governments achieve greater control over fishing activities. At the international level, a number of new agreements have created a stronger legal basis on which to develop greater control. At the same time, new technological developments have facilitated the remote monitoring of fishing vessels and the collection of fisheries data.

The 1982 United Nations Convention on the Law of the Sea (UNCLOS), which entered into force in 1994, forms the backbone of the international legal framework for fisheries management. It sets out the rights and duties of coastal, port and flag States in respect of each of the principal maritime zones recognized by international law, namely the territorial sea, the exclusive economic zone and the high seas. It also deals with a range of other important issues that are related, including the legal regimes applicable to internal waters, archipelagic waters, the contiguous zone, the continental shelf, and the right of innocent passage and passage through international straits. The provisions of UNCLOS relating to fisheries, though widely accepted even before the Convention entered into force, did not prevent the depletion of several valuable fish stocks. For this reason, the 1992 United Nations Conference on Environment and Development (UNCED) called urgently for the development of further instruments that would be necessary to re-establish and maintain sustainable fisheries worldwide.

One of these new instruments is the Agreement for the Implementation of the Provisions of the United Nations Convention on the Law of the Sea of 10 December 1982 Relating to the Conservation and Management of Straddling Fish Stocks and Highly Migratory Fish Stocks. The UN Fish Stocks Agreement sets forth a broad range of obligations designed to create greater control over fisheries for certain valuable stocks, including the strengthening of MCS capabilities.

The pertinent aspects of several other new instruments developed under the auspices of FAO, including:

- The 1993 Agreement to Promote Compliance with International Conservation and Management Measures by Fishing Vessels on the High Seas;
- The 1995 Code of Conduct for Responsible Fisheries (CCRF);
- Four International Plans of Action dealing with various aspects of fisheries management, and particularly the International Plan of Action to Prevent, Deter and Eliminate Illegal, Unreported and Unregulated Fishing;
- Guidelines for the marking and identification of fishing vessels.

Ideally, each State would implement the 1982 Convention and the more recent agreements through the development of a national "oceans policy." Such a policy would establish government priorities and the strategy for the conservation and sustainable use of all marine resources within the maritime zones over which the State exercises sovereignty or sovereign rights, as well as efforts towards cooperative management of fisheries that occur outside these zones. From this oceans policy would flow the integrated oceans planning and management framework under which fisheries management plans would be developed. Most States see this as a long-term development initiative. To shorten the process, they have chosen to develop oceans policy and fisheries management strategies (including MCS strategies) simultaneously. Consequently, although States recognize that fisheries management must be integrated into an overall oceans policy when it is ultimately established, the MCS systems required to implement fisheries management plans are being developed in the interim to address the more immediate need to protect fish stocks and their habitats. This strategy is commendable. A "precautionary approach" recognizes that a first requirement for fisheries resource conservation is to prevent further degradation of the resource base.

The degree to which a government becomes involved in the fishing industry will have an impact on fisheries management and the resultant MCS activities.

For example, a government can:

- Assume a controlling role, where it actually runs the industry, impacts the potential income of fishers, and micro-regulates the harvesting sector; or
- Maintain a less intrusive or co-management role, whereby the fishers and the fishing industry are encouraged to accept their resource responsibilities and roles within the framework of general government conservation principles and legislation.

Negative results of centralised, micro-management control mechanisms have become evident in both industrialised and developing States. Consequently, there is an emerging trend towards the second, *participatory co-management approach.* Fishers and the fishing industry want, and in fact are demanding, a more active role in management planning and implementation. Central governments are responding by devolving authority to smaller units of government (provinces, districts and municipalities) and by fostering community-based management, stakeholder involvement, and the acceptance of responsibility for the care, conservation and protection of their local marine resources by the fishers and industry. In a growing number of cases, such as the Canadian experience described in Profile 1, the private sector is also becoming involved in MCS activities.

MISPERCEPTIONS OF MCS

Misperceptions surrounding MCS activities continue to impact how fishers and even fishery managers view the process of MCS and enforcement. A common misperception is that all fisheries problems stem either from a failure to control illegal foreign fishing, or from the fishers themselves. While foreign fishing fleets have had documented impacts on fisheries conservation efforts, the greater impact on fisheries often stems from the domestic fishing industry in the coastal and nearshore fishing zones. On the issue of fishers, fisheries administrators must remember that most are hard working individuals often working in a hazardous environment:

While sometimes libellously assumed by the ill-informed to be crooks, [fishers] are perhaps best described as being as honest as the next man, but hard, individualistic businessmen running very competitive and often highly capitalized operations. It is worth remembering that they do so in the face of a largely unforgiving sea that creates a working environment which... [has one of the worst industrial accident rates in the world], and [that]... they operate increasingly in an economic climate of ever increasing overheads countered only by the proceeds of catches which [are] subjected to [ever greater] quota restriction. All of this is done in the knowledge that the success of their venture and the livelihood of their crews depends entirely on their individual skill, effort and initiative. Given these pressures, it is perhaps not surprising that such independent minds do not always take kindly to bureaucratic controls, especially if these appear to them to have little practical purpose. Another erroneous perception is that MCS is exclusively concerned with enforcement - thus ignoring the other two components of monitoring (data collection) and control (legislation, licensing, and controls on gear, season, areas, etc.). In focussing only on the "surveillance/enforcement" or deterrent aspects of MCS, fisheries administrators and supporting agencies cannot harness the full utility of MCS as the vital executive arm of fisheries management.

Civilian versus Military Involvement in MCS

The expense of MCS activities is often a primary concern of any government designing and implementing an MCS system. Cost-effectiveness and efficiency is important if MCS operations are to be successful. A civilian approach to deterrent fisheries enforcement has proven in many cases to be the most cost-effective and responsive to fisheries priorities. Use of civilian assets also minimizes the political sensitivity of international fisheries incidents by avoiding the use of military equipment and personnel. Those fisheries administrators who must rely on the use of military resources to carry out MCS activities may find that military agencies often accord low priority to that task.

Moreover, military involvement, except in a support role, is usually not cost-effective. Military aircraft and vessels are more expensive to build and operate than equivalent civilian equipment. Savings accrue from the use of a civilian vessel with fewer crew, and lower operating costs. For many governments, however, the military can play a significant supporting role in a strong MCS system. The key for such governments is to establish an inter-agency mechanism that enables fisheries administrators to call upon their military counterparts as and when needed.

Fisheries as a Lead Ministry

Effectiveness of operations can be enhanced considerably if a single ministry is designated to take the lead role in MCS activities. This significantly reduces the lines of communications for the command and control of the monitoring and surveillance components of MCS activities, making them more efficient and responsive to management needs. As noted above, however, a number of different agencies may be called upon in a supporting role. In such situations, effective MCS requires a strong inter-agency control mechanism.

In Canada, the federal Department of Fisheries and Oceans (DFO) has been entrusted by the Parliament of Canada through the *Fisheries Act* and the *Coastal Fisheries Protection Act* to administer all laws relating to fisheries. The administration of federal fisheries laws has, by agreement, been delegated to some, but not all, provincial governments. DFO remains responsible for fisheries management in the tidal waters of the Pacific, Atlantic, Arctic, the inland waters of four Atlantic provinces and the salmon rivers in British

Columbia. This includes management of Aboriginal, Recreational and Commercial fisheries within Canada's Exclusive Economic Zone (EEZ), in transboundary rivers, and for sedentary species on the continental shelf outside the Canadian EEZ.

Statistics:

- 58 400 commercial fishers (42 700 Atlantic, 8 700 Pacific and 7 000 inland)
- Commercial harvest > 1 000 000 mt
- Landed value approximately $1.9 billion (shellfish account for approximately 75% of the total landed value, with groundfish and pelagic fisheries making up the remainder)
- 5 million Canadians and 900 000 visitors/year for the recreational fishery
- Catch over 250 million fish in recreational fisheries, with more than half of the fish released under the catch and release regime
- Over 125 fisheries agreements are negotiated with Aboriginal groups in Canada on an annual basis to provide for aboriginal access to fisheries and an orderly management of their fishing activities.

Management Systems/Control Mechanisms

Various management schemes have been applied in different fisheries, but all are based on limited entry licensing, with vessel and gear restrictions. Other measures to limit catches include total allowable catches (*e.g.* groundfish), escapement targets (salmon), or recruitment strategies (*e.g.* lobster). Other management measures include limitations on fishing area, fishing season, gear (*e.g.* mesh size), incidental catch (bycatch) and minimum fish sizes. Rights-based systems - in the form of Enterprise Allocations (EA) and Individual Quotas (IQ) - have been introduced in some fisheries to allow fishers more efficiently to manage the capacity and effort for harvesting.

MCS Programme

The Department of Fisheries and Oceans (DFO) Conservation and Protection programme ensures compliance with the legislation, regulations and fishing plans. This requires an integrated MCS approach and deployment of some 600 Fishery Officers for air, sea and land patrols; independent/private sector observer coverage on fishing vessels; dockside monitoring of fish landings; and remote electronic monitoring of fishing vessel activity.

DFO operates a fleet of patrol vessels on each coast to enforce closed areas and boundary lines, and to conduct inspections at sea for compliance purposes. Contracted aircraft are used to monitor fishing fleets. Sea and aerial surveillance is also supplemented by the Department of National Defence (DND). VMS and air surveillance provide for more effective deployment of patrol vessels; with the latter also serving as a visible deterrence. Canada deploys private sector, contracted observers, without enforcement powers, on all foreign vessels fishing in Canadian waters and on some Canadian vessels to gather scientific information and provide on-site monitoring of compliance. They are trained to detect and report infractions such as dumping/discarding, fishing in closed areas, catch misreporting, retention of prohibited catch and the use of illegal gear. The level of observer coverage in domestic fisheries varies depending on conservation risks and management priorities. Contracted dockside monitors/ observers verify the quantity and species/product form of fish landed for scientific, quota monitoring and compliance reasons. This data is cross-checked by random inspections by Fishery Officers at landing sites. At-sea observer costs are shared by the Department and the fishing industry. Dockside monitoring costs are entirely paid by the fishing industry. In 1999, DFOs expenditures on fisheries enforcement amounted to CDN $70 million (including DFOs share of at-sea observer costs).

The Conservation and Protection programme has been significantly re-oriented in recent years. The mix of enforcement resources has been altered to better respond to changing programme requirements. For example, a number of larger patrol vessels have been replaced with smaller programme boats that can be operated more efficiently by the Fishery Officers themselves. Savings from vessel reductions have been partially re-invested in new equipment and surveillance technologies. Significant investments have also been made in the creation of new enforcement data systems and the integration of existing systems, with the goal of providing Fishery Officers and managers with more accurate and timely information that will strengthen the Department's enforcement capabilities. These efforts at improving data integration and analysis will continue to be a priority for the immediate future. The Conservation and Protection programme is closely integrated with DFOs overall Fisheries Management Programme. Input and advice from fisheries enforcement officials is an important consideration in the development of Integrated Fisheries Management Plans (IFMPs). In future these plans will include specific conservation objectives as identified by DFO scientists in consultation with fishers and other technical experts. The new Objectives Based Fisheries Management strategy will require greater involvement by the fishing industry to design management measures that will minimize identified risks to conservation.

Conserve and Utilise Marine resources

Fisheries are critical to the development of a State's plan to conserve and utilise marine resources, as fish and their habitat are significant renewable resources in the territorial sea and exclusive economic zone. The goal of fisheries management, including MCS, is to maximise the economic opportunities and benefits from the State's waters within sustainable harvesting limits. Fisheries MCS needs to be defined in light of this goal. An MCS

Conference of Experts organized by FAO in 1981 developed a definition of MCS that is commonly accepted by fisheries personnel:

- Monitoring - the continuous requirement for the measurement of fishing effort characteristics and resource yields;
- Control - the regulatory conditions under which the exploitation of the resource may be conducted; and
- Surveillance - the degree and types of observations required to maintain compliance with the regulatory controls imposed on fishing activities.

Simply stated, MCS is the mechanism for implementation of agreed policies, plans or strategies for oceans and fisheries management. MCS is an aspect of oceans and fisheries management that is often undervalued. In reality, it is key to the successful implementation of any planning strategy. The absence of MCS operations render a fisheries management scheme incomplete and ineffective. Since the 1981 MCS Conference, the definition of MCS has been enhanced to promote the concept that MCS covers more than just fisheries enforcement - it is an integral and key component for the implementation of fisheries management plans. It encompasses not only traditional enforcement activities but also the development and establishment of both data collection systems, the enactment of legislative instruments and the implementation of the management plan through participatory techniques and strategies.

A 1993 workshop in Ghana offered the following clarifications:

- *Monitoring* includes the collection, measurement and analysis of fishing activity including, but not limited to: catch, species composition, fishing effort, bycatch, discards, area of operations, etc. This information is primary data that fisheries managers use to arrive at management decisions. If this information is unavailable, inaccurate or incomplete, managers will be handicapped in developing and implementing management measures.
- *Control* involves the specification of the terms and conditions under which resources can be harvested. These specifications are normally contained in national fisheries legislation and other arrangements that might be nationally, subregionally, or regionally agreed. The legislation provides the basis for which fisheries management arrangements, via MCS, are implemented. For maximum effect, framework legislation should clearly state the management measures being implemented and define the requirements and prohibitions that will be enforced.
- *Surveillance* involves the regulation and supervision of fishing activity to ensure that national legislation and terms, conditions of access, and management measures are observed. *This activity is critical to ensure that resources are not over exploited, poaching is minimized and management arrangements are implemented.*

These wider definitions amplify the importance of all aspects of MCS.

EMERGING TRENDS IN MCS

Trends over in the past decade have significantly altered fisheries management in general and MCS strategies in particular. These have occurred in three areas - viz: *devolution of authority to lower levels of government; the encouragement of participatory management*; and *new technology*.

Devolution of Authority

Fisheries management can no longer be the sole responsibility of central governments. Rapid population growth, urbanisation and migration to coastal areas have created a greater need for local governmental units to address local fisheries management concerns. Growing literacy and educational levels, coupled with easy access to mass media, have helped ordinary citizens to acquire more knowledge of government processes and a desire for more transparency and a greater voice in decision-making. Central governments have responded by devolving more authority over fisheries management to provinces, states, districts, municipalities and communities.

This devolution of authority has vitally affected MCS. In the past, central governments controlled, or at least sought to control, all MCS activities on a national basis. Today, a more typical role for a central government is to set national policies and standards and to co-ordinate the implementation of these policies and standards. Responsibility for actually carrying out MCS activities is now typically shared by central governments with smaller governmental units. Central governments also monitor and evaluate local government performance to enhance future coordination and planning. MCS planning and operations still need this national coordination for consistency in implementation of management plans, and for conflict resolution. However, the overall trend is for central governments to devolve responsibility for MCS activities to lower levels of governmental authority.

Namibia, a State six times the size of the United Kingdom and with a population of 1.7 million, is on the West Coast of Africa just north of South Africa. It has a coastline of 700 nautical miles (1 500 km) and an EEZ of 275 000 km^2. Since independence in 1989, Namibia has developed a progressive fisheries administration and a thriving commercial capture fisheries that is increasingly Namibian.

The fishing industry is based on the high productivity of the up-welling Benguela Current system. Fisheries contributed more than US$50 million (10% Namibia's GDP) in 1999 and employed some 15 000 people. By law, all fish are landed in Namibia through its two ports, Walvis Bay and Lüderitz. Namibia is one of the few major fishing nations that earns more income and creates more jobs from the processing sector than from the catching of fish. This is partly due to policies such as the requirement for onshore processing of hake and other species. *Namibia is recovering all expenses related to its fisheries administration and MCS operations through the collection of revenue from the fishing industry.*

The fisheries are exclusively industrial and annual catches are stable or steadily increasing following the overexploitation before independence. Trends for the three major species (hake, horse mackerel and pilchard) are showing the same broadly consistent pattern of increases in biomass up to 1992, a subsequent decline and, recently, a general recovery. The total landed catch in 1999 was approximately 600 000 t.

The key fisheries are as follows (TAC for 2000 in brackets):

- *The Demersal Fisheries*: Hake (200 000 t), monk (16 000 t), kingklip and sole;
- *The Midwater Fishery*: Horse mackerel (410 000 t) pursued by both midwater trawlers and the purse seiners;
- The Purse Seine Fishery: Pilchard (15 000 t) and juvenile horse mackerel with purse seine nets;
- *The Deep-Water Fishery*: Orange roughy (2 400 t) and alfonsino;
- *The Large Pelagic Fishery*: Albacore, big-eye, yellow-fin, skipjack and swordfish;
- *The Rock Lobster Fishery (400 t)*: Small vessels using carrier vessels to bring the live lobster ashore daily;
- *The Crab Fishery (2 000 t)*: A small fishery that uses traps to catch deep-sea crab during the whole year.

Namibia uses a three-part access system to turn its policies of Namibianisation of the fisheries into practise:

i. Rights to exploit the fisheries are allocated;
ii. Quotas are allocated to rights holders; and
iii. Vessels are granted licences.

This allows Namibia to maintain control over the fishing companies, the vessels and the crews in order to assure that policy aims are being met through a differentiated quota fees system that favours Namibian participation, Namibian flag vessels, Namibian employment, and an empowerment aspect in relation to the social inequalities.

Fisheries management controls fall into two categories:

- Input controls that relate to fishing effort and gear, seasons and areas, with the key ones being the limitation of total fishing effort and seasons; and
- Output controls that set quotas and regulations, size limits, and other characteristics of the fish that may be landed. The main output control is by the establishment of annual TAC and quota allocations set on the best scientific advice available, modified by socio-economic factors. TACs are today established for 7 species (hake, horse mackerel, pilchard, orange roughy, monk, red crab and rock lobster).

The MCS system is an integrated system that has stations in the two Inspectorates in Walvis Bay and Lüderitz with each station tasked as appropriate for deploying fishery officers to air, sea or land operations, deploying fisheries observers on board fishing vessels, analysing past operations and outputs or planning future operations. There is a cross-verification of data where observer and inspector information is checked through port inspections and landings control. MCS activities related to air and sea operations are co-ordinated from the Inspectorate at Walvis Bay through a maritime and fisheries Operations Centre and an air base at Arandis.

The Ministry of Fisheries and Marine Resources operates two larger patrol vessels and one fixed wing aircraft in addition to maintaining 100% observer coverage of the fishing fleet (230 observers are employed in the largest observer programme south of equator). Observers and fisheries inspectors control gear and catches at sea. All catches are finally landed or transhipped in Namibia under the supervision and control of fisheries inspectors. VMS is still in a pilot stage and different systems are under evaluation.

Participatory Management

Similarly, governments at all levels are giving private groups and individuals more opportunity to participate in the management of fisheries that affect them. This trend towards more transparency and openness is changing all aspects of MCS, including the development of fisheries laws and regulations, the collection of fisheries data, etc. As already indicated, enhanced public participation in the fisheries management process brings a number of advantages.

Specifically, it fosters:

- Greater public understanding of the rationale behind MCS activities;
- A greater sense of partnership between the government and fishers and others whose activities are being regulated;
- A greater sense of ownership and acceptance by private groups and individuals in the decisions that are ultimately adopted;
- Voluntary compliance and the use of peer pressure to deter violations of fisheries rules;
- Greater availability of MCS funding and other assets for dealing with cases of non-compliance.

New Technology

Advances in technology, along with reductions in the cost of technology, have revolutionised MCS.

Examples include:

- New vessel monitoring systems (VMS) in use in Argentina, Australia, Canada, Europe, Malaysia, New Zealand, United States, and member States of the South Pacific Forum Fisheries Agency, with systems also being considered in Madagascar, Namibia, Indonesia, and Sri Lanka;

- Inter-linked coastal radar systems in use in Senegal; and
- A combined VMS and satellite imaging system in use in the Maldives.

Cheap cellular telephone and computer technology has also brought MCS into the computer age and has greatly improved the capacity of States to respond rapidly to situations. Over-the-horizon radar technology is evolving as well, but its cost has not yet fallen to a level that is appropriate for general fisheries management.

Currently the most popular VMS systems use INMARSAT C (and more recently D) and ARGOS. Others such as POLESTAR and EULTRACS are also making an appearance. FAO has prepared detailed guidelines as part of the *Code of Conduct for Responsible Fisheries* series for Fisheries Administrators contemplating the introduction and use of such technology.

The use of this new technology for coastal fisheries, particularly VMS, may not yet be cost-effective. However, many States can still monitor large numbers of coastal vessels effectively through other means, *e.g.* observers, dockside monitoring, monitoring of industry landings and production records, etc.

ROLE OF MCS IN FISHERIES MANAGEMENT

Fisheries management in its simplest terms comprise the following activities:

- *Data collection and analysis*: Data for management planning and operations from socio-economic studies, rural development studies, fisheries population studies, fisheries research cruises, licensing (national, provincial and district), catch and effort/logbooks, onboard observers (if established as a programme), dockside monitoring/ landings, VMS, satellite imaging, inspections at sea and in port, etc.
- *Participatory management planning*: Planning of fisheries management policies and strategies at the national level, and detailed planning for management zones or areas with input from stakeholders (provinces, districts and fishers).
- *Establishing a regulatory framework:* The management plans need to be supported by appropriate legal instruments by means of which the plans are implemented. These legal instruments detail all the control mechanisms available for fisheries management including, but not limited to:
  - *Input controls:* Such as access (number of fishers, number of vessels by fishery), licences, closed seasons, gear restrictions, vessel limitations, area restrictions (Protected Areas), VMS requirements, and vessel identification.
  - *Operational and output controls*: Such as species and catch limits, by-catch limits, reporting requirements, air surveillance, sea patrols/inspections, boarding, logbooks, dockside monitoring, observers, port inspections, and catch documentation schemes.

- *Implementation*: This includes such measures as:
  - Participatory community-based management (CBM);
  - "Preventive" MCS activities to encourage voluntary compliance;
  - Public awareness and education campaigns;
  - Assistance to small scale fishers for supplemental livelihood development to reduce coastal area pressures;
  - Full enforcement to ensure compliance by those minority of fishers that persist in ignoring the law.

Similarly, MCS involves:

- Data collection and analyses for both operational planning and execution, as well as management planning, defined as monitoring (M);
- Involvement in the participatory management planning to include discussions on appropriate implementing mechanisms;
- Development and approval of appropriate and enforceable legislative instruments and control mechanisms such as licences, permitted fishing gear, seasons, vessel sizes, fish sizes, species, catch limits, by-catch limits, and area controls, or other restrictions to support the management plan, referred to as the control mechanisms (C); and
- Implementation of the plan through "preventive" and "deterrent" MCS techniques, included in the idea of surveillance (S).

Unfortunately, not all fisheries administrators understand MCS, or its critical role as an implementing mechanism for fisheries management. Some view arrests as the only relevant indication of the effectiveness of MCS efforts.

The real indicator for MCS is the level of compliance, and this is governed by many factors, *e.g.* the number of fishers; the number of vessels; effort and area coverage of patrols; results of patrols, increase in voluntary compliance, etc.

Effective MCS involves a two-pronged, parallel approach. The *preventive approach* is to encourage "voluntary compliance" through understanding and support for the management strategies and this includes:

- Enhancement of community/fisher awareness and understanding of management practices and MCS through seminars, public awareness and information, education, and communication campaigns;
- Participatory management development to promote ownership of the management regime and input into the regulatory/control aspect of management (laws and regulations) in preparation for acceptance by the fishers of their joint "stewardship" role for the management of their fisheries in partnership with government;
- Peer pressure towards voluntary compliance and support for the management regime;
- The institution of accurate and verifiable data collection regimes; and
- Surveillance and verification for compliance.

The parallel approach of *deterrent/enforcement* MCS is necessary to ensure compliance by fishers who resist the regulatory regime to the detriment of both the fishery and the economic returns to their fellow fishers. Deterrent and enforcement include inspection, investigation, prevention and court proceedings to enforce the law. Voluntary compliance will fail if stakeholders see non-compliant fishers successfully evading the law and receiving economic returns from their illegal activity, at the expense of the fishers who comply with all requirements.

Namibia has used an effective mix of input and output controls to build a very successful fisheries management regime in the short period since independence was declared in 1989.

MCS SPATIAL COMPONENTS

There are three main spatial dimensions or components to MCS: land, sea and air. The proper configuration of these spatial components for a given system will depend on such factors as cost, commitment, and organizational structure (national, subregional, or regional). All three components can now effectively use satellite technology (*e.g.* for vessel licensing, data collection and enforcement operations).

The *land* component of an MCS system serves as the base of operations, the co-ordinating centre for all MCS activities from which governments can regulate the deployment of resources to best address changing situations. The land component also entails port inspections, dockside monitoring, and the monitoring of transshipments and trade in fish products to ensure compliance with relevant rules.

Governments must also undertake a variety of land-based activities in order to carry out their responsibilities as flag States of vessels that may be fishing in remote areas, including on the high seas and in waters under the jurisdiction of other States. New technology has allowed States to link the land components of their MCS system to those of other States on a regional or subregional basis, which can greatly foster co-ordinated and responsible management.

The *sea* component includes MCS activities undertaken in marine areas under the jurisdiction of a State and may also cover high seas areas. Technology that comes into play can include radar, sonar and vessel platforms. While physical presence through at-sea patrols remains a fundamental part of this component (necessary for arresting violators and securing evidence), the costs involved have prompted a growing number of States to employ "no force" surveillance techniques. Examples of such techniques include the placement of independent observers on board fishing vessels, national or regional vessel registers and VMS requirements. Developing coastal States are also requiring the flag States of vessels that wish to fish in their EEZs to ensure that those vessels do not violate the terms of the access granted. However, an at-sea presence remains necessary for the active enforcement of management measures to protect marine fisheries resources.

The *air* component covers the air and space equipment (aircraft, satellites, etc.) used in MCS activities. The flexibility, speed and deterrence of air and satellite based surveillance systems make these very popular tools for fisheries management. The air component provides for the rapid collection and dissemination of a wide range of information, including fishing vessel identification and reported fisheries data. Air, satellite or VMS surveillance can often provide initial information regarding fishing activity; they can also serve as first indicators of potential illegal activity and can thus trigger further MCS action. The cost of the air component is directly related to the sophistication of the technology used. Because governments can use air, satellite, video and advanced digital photographic technology, and VMS surveillance technology to address tasks beyond MCS (*e.g.* environmental and coastal zone monitoring, customs, immigration, and search and rescue), this equipment can improve the cost-effectiveness and efficiency of an integrated ocean management programme.

The United States of America has used all these tools in the past and currently has a multi-agency approach to MCS as noted.

The United States EEZ is reputedly the largest in the world at 3.36 million square miles (8.684 million km?). It hosts some 100 000 commercial fishing vessels. The 1999 fisheries resulted in landings of 9,339,034,000 pounds (4.2451 million metric tonnes) valued at US$ $3.4671 billion. Pollock, menhaden and salmon in that order accounted for 40% of the landings of the top ten species; while shrimp, crab and salmon brought in the highest revenues.

The management of fisheries is shared with the federal government and the states with the latter having management authority to 3 nm. The 200 mile EEZ was declared in the late 1970s at which time the USA permitted foreign fishing vessels in their waters under very strict conditions of entry and exit reporting, licensing, observers, and access only to specified windows for fishing with further gear, catch and effort controls. Foreign fishing ceased in USA's waters in the early 1990s, as the United States fishing fleet increasingly encountered pressures of over capitalization similar to other States. MCS control mechanisms for the domestic fisheries include the use of: licenses, quota, season, gear and area controls with appropriate reports and landing verifications, and observers for specific fisheries.

Vessel monitoring system technology is in use in the USA for fisheries vessel tracking on approximately 500 vessels in Hawaii, New England and Alaska, with plans for expansion for another possible 3,330 vessels in the Atlantic and Gulf of Mexico.

MCS Agency Responsibilities:

Each individual State is responsible for its 3 nm zone. The National Marine Fisheries Service (NMFS) under the National Oceanic and Atmospheric

Administration (NOAA) conducts shore monitoring and at times places personnel aboard multi-agency assets for MCS, *e.g.* air patrols and sea patrols. The United States Coast Guard (USCG) is the key agency responsible for enforcement within the EEZ and their air and sea assets are multi-tasked with fisheries being one of the secondary tasks. Coast Guard vessels spent 92 864 hours in fisheries related operations in 1999. USCG air hours for fisheries for 1999 were 10,527 hours. (This does not include the individual State efforts and dedicated efforts for MCS in their coastal waters.) The NOAA Office for Law Enforcement has five divisions based out of Silver Springs, Maryland and supports 50 field offices throughout the United States and Territories. It is the acknowledged leader and expert in the field of marine resource enforcement in the United States. The United States is developing an integrated national system so all fisheries that may be subjected to VMS requirements will be centralized for analysis, planning, and operations.

2

Challenging Role of Fisheries Sector

FISHERIES SECTOR

Fisheries sector can play a vital role, as a potential source, in attaining nutritional security in India. The current production of fish in India forms only 71.0% of the total potential and hence there is an ample scope of improvement and thus can be added up to the nutritional security. It is estimated that globally, about a third of the fish catch is not utilized for human food consumption because of post harvest loss resulting from poor handling and preservation. In India, there is also the problem of under utilization of by- catch due to the many species that are caught in the net. The cost effective and efficient utilization of aquatic products demands proper processing and distribution. Demand for fish and fish related products are increasing day by day in our country and reduction in post harvest losses can make a major contribution to satisfying this demand, improving quality and quantity for consumers and increasing income for the producers.

HARVEST UTILIZATION OF FISH CATCHES

Thus, improvement in the post harvest utilization of fish catches can ensure further nutritional security among a wide range of people in our country. Viewing fish primarily as a source of protein has been the dominant perspective in studies on nutritional security. This view is primarily not in the right direction, since it fails to highlight the critical role of fish as a wholesome source of Poly Unsaturated Fatty Acids (PUFA's); minerals such as calcium, phosphorus, iron; vitamins like A, B_1, B_2, B_{12}, D etc.; and trace elements such as iodine and zinc. These attributes makes fish a vital contributor to nutritional security of the most deprived and vulnerable populations. For them, easily digestible fish is indispensable; other natural animal or vegetable protein sources are poor substitutes on both nutritional as well as economic grounds.

ROLE OF TIFAC IN FISHERIES SECTOR

Technology Information, Forecasting & Assessment Council (TIFAC) is an autonomous society under the Department of Science & Technology (DST), Govt. of India. Following the Technology Policy Statement of 1983 and the Technology Policy Implementation Committee recommendations, TIFAC was set up in 1988 for keeping technology watch on priority areas and to promote actions. A landmark achievement of TIFAC has been the first major national long-term technology forecasting exercise known as Technology Vision for India up to 2020. Major thrust of TIFAC activities includes undertaking selected projects in identified sectors among which Agriculture and Agro Food Processing forms a major component.

AGRO FOOD PROCESSING SECTOR

Under the Agro Food Processing sector, the fisheries sub sector was constituted under the chairmanship of Dr. S.A.H. Abidi, Member, Agricultural Scientists Recruitment Board (ASRB), Indian Council of Agriculture (ICAR), New Delhi, and few other national fisheries experts as members, with a view of transforming the cottage level fisheries industry, by:

(a) establishing commercial level plants, with high replication potential, for processing/ producing quality fish products to capture enhanced domestic and foreign market and

(b) creating job opportunities for local community utilizing the local resources through science and technology inputs.

This chapter reviews the success stories of TIFAC fisheries projects on development of value added fish products as well its future targets.

Socio-economic Development

To ensure rapid socio-economic development, application of science and technology should be the foremost priority. In this background, TIFAC has initiated various projects on the development of value added aquatic products in different geographical areas of our country with different women SHG's/ Cooperatives in collaboration with fisheries institutes as well as state fisheries departments; who are providing the technical and logistical support to the projects. The projects are selected and approved by the Fishery Panel based on the technical as well as economic feasibility of the projects, as recommended by the experts in the concerned areas. A Project Review and Monitoring Committee (PRMC), consisting of experts nominated by TIFAC, are constituted for each project for reviewing and monitoring the progress made by each project. The PRMC will review the progress relating to the project and provides technical advise from time to time at various stages of completion. The PRMC after closely monitoring the developments in the projects may provide mid course remedial actions, if required, to overcome the unforeseen obstacles.

COMMERCIAL LEVEL COTTAGE INDUSTRY

The objectives of the projects are to establish a commercial level cottage industry by organizing women fishers and to improve upon their socio-economic status; to improve the utilization of low value fish/ shellfish by value added product development; improvement in quality & quality testing of the product; and market linkage.

Fish Processing and Preservation

The projects have facilitated and efficiently transferred the technologies such as fish processing, preservation, packing, quality control and marketing as developed by the Fish Processing Laboratory of CIFE. This was done through training programmes. The training and exposure imparted during the programmes enabled the society to address various product quality related issues through the implementation of quality assurance techniques. The advantages of the technology adopted include superior quality, greater shelf life, and higher acceptability. Various value added fish products were prepared from low cost fishes. The products include:

1. *Prawn Pickles*: Pickling is one of the safest means of easy preservation of fish/ shell fish. Pickles are good appetizers and they add palatability to the starch based blend to taste Asian dishes, besides being highly nutritious. The technology involved is simple and can be imbibed by rural women with great ease. A new recipe has been developed, by CIFE, to improve the acceptability and the shelf life. Novel methods of preparations were used to get pickle with good eating quality.
2. *Fish Papad*: Papad is a direct product commonly preferred all over India as a side dish. Fish is incorporated in papad to make papad more nutritious and tastier. The flour of black gram pulse (Urad dal) is used as the major ingredient. Fish was further added to make papad more nutritious and tastier.
3. *Fish Chakli*: Piston or ram type extruder is used in different regions of India to prepare starch or pulse based fried snacks. Chakli (spiral) is one such popular product in Maharashtra. Fish meat is incorporated in chakli to enhance its taste, flavor and nutritive value.
4. *Fish Sandwich*: Using minced fish meat a product was prepared, which could be spread between bread pieces to make fish sandwich.
5. *Fish Paste*: The product is prepared by using fish flesh (minced) with spices and preservatives; which can be used as a side dish along with rice or bread having a shelf life of three months.
6. *Fish Noodles*: Fish flesh is made in the form of paste and then mixed with refined flour to produce delicious fish noodles.
7. *Fish Puff*: By using twin screw extruder a directly eatable cooked product containing ingredients like fish flesh, rice and gram powder with salt and spices can be prepared with a shelf life of six months.

Small Scale Commercial Cottage Industry

The projects have successfully demonstrated the potential for processing low value fish efficiently and incorporating them in acceptable, nutritious and safe, novel and traditional products in a small scale commercial cottage industry.

The projects enabled the poor fisherwoman to organize and to form SHG's for attaining sustainable livelihood. Training programmes were useful in respect to give SHG's an insight into modalities of organizing product processing and marketing. The project helped to create awareness on the potential for value-addition and market diversification among a wide segment of the local people, thus facilitating the technology to get replicated in other parts of the country. The fishery resources have been utilized maximum leading to reduction of post harvest losses.

Due to the higher value of value added products, when compared to low cost fish has earned the women group a handsome profit thus improving their socio-economic status. Stringent quality checking by CIFE resulted in production of value added products with superior quality, thus enhancing the shelf life and profit. The projects facilitated closer interaction between local traders through visits to major markets and trade fair participation. Continued buyer-seller matching, and provision of market information assisted in market development. At the end of the project period, the societies have shown an improved access to major markets through expanding their value added product range.

Value Added Fish Products

The success of projects in the area of value added fish products from low cost fishes have prompted TIFAC to replicate it in other areas and also to initiate projects in other areas also. The major thrust areas which TIFAC is focusing for its future activities include breeding and culture of ornamental fishes, production of sheedal; and caviar and fillet production from trouts.

1. *Ornamental fish breeding and culture*: Ornamental fish keeping and its propagation has become an interesting activity for many, providing not only aesthetic pleasure but also financial openings. It has reached new heights in the recent past and has gained accelerated momentum, thus providing employment opportunities to many including unemployed youth, women and other unskilled masses. Indian waters possess a rich diversity of ornamental fishes with over 100 varieties of indigenous species, in addition to a similar number of exotic species that are bred in captivity. The strategies to be adopted for boosting commercial production of ornamental fishes includes technical support and dissemination of recent research developments in breeding and culture; formulation and preparation of nutritious feed;

health management; water quality management; and training programme for creating scientifically and technically skilled man power

2. *Sheedal Production*: Sheedal is a form of traditionally fermented fish product prepared in the North-eastern region of India. Apart from its delicacy and food value, people of this region like sheedal very much because of its medicinal value for preventing stomach related disorders and malaria. The process of preparation varies from place to place and community to community. *Puntius spp.* is the major species of fish used for sheedal preparation. For sustainable production and uninterrupted marketing of sheedal the approach involves training, mass production, quality up-gradation, and market development strategies.
3. *Caviar and Fillet production from trout*: Caviar and fillets are premium products which have high demand and fetches greater value in the international market. Caviar is produced from the roe of trout and filleting provides quality trout flesh. Trouts are fresh water fishes seen in cooler climate and high altitudes. Jammu & Kashmir is blessed with trout resource and has been identified as the ideal place to launch such a project.

For initiating a project on caviar and fillet production, the major identified inputs includes, sufficient raw material, sophisticated processing equipments, trained man power to operate the project, and recent technologies and research developments. Aquatic resource is a vital tool for transforming India into a nutritionally secured nation. Development of value added aquatic products can bring both improvement of the socioeconomic status as well nutritional security in our country. The efforts of TIFAC in this direction have shown fruitful from the encouraging results obtaining from various projects. By producing quality value added products and establishing the market, the SHG,s can help India to march towards prosperity as well as self sufficiency.

INTEGRATED FISHERIES PROJECT (IFP)

Established as Indo Norwegian Project during 1952 this was renamed as Integrated Fisheries Project (IFP) in 1972. The Project is engaged in the development of technologies for harvesting and post-harvesting of marine fish resources. It has a well established Fishery complex consisting of a fishing fleet, a modern mechanical workshop and Slipway to slip vessels up to a displacement weight of 250 tonnes, an Ice cum freezing Plant, a well equipped processing unit for processing and marketing of diversified value added fish products, a research and development laboratory to carry out quality analysis of raw fish, finished products handled by the Project and testing and developing suitable packages for different value added products and Life Raft Servicing Station. The Institute has a unit at Vizag also.

Mandate of the Project

(a) Technology development and transfer to beneficiaries consisting of rural fishermen community, small scale industries and export processing houses through consultancy and job work.
(b) Value added product development by way of process and product diversification from all varieties of fish including low value and unconventional species.
(c) Imparting training in the field of Post Harvest Technology, Refrigeration Technology, Quality control and Value added Products.
(d) Providing consultancy and training for rural development programmes like supporting local fish farmers, self-help groups of fisher community and fishermen's co-operative societies functioning and Panchayati Raj Institutions. And also to Women empowerment programmes in processing and establishment of SSI units to self help groups.
(e) Popularization and test marketing of value added products from all fish varieties including low value and unconventional species.
(f) Extension of the popularization and test marketing of value added products to new areas and developing market in all states in a phased manner with added attention to rural areas enthusing entrepreneurs to enter into seafood processing industry.
(g) Function as a nodal agency on all matters associated with quality assurance, standardization of fish products and safety of fish products.

POSTMORTEM CHANGES IN FISHERIES SECTOR

SENSORY CHANGES

Sensory changes are those perceived with the senses, *i.e.*, appearance, odour, texture and taste.

Changes in Raw Fresh Fish

The first sensory changes of fish during storage are concerned with appearance and texture. The characteristic taste of the species is normally developed the first couple of days during storage in ice. The most dramatic change is onset of rigor mortis. Immediately after death the muscle is totally relaxed and the limp elastic texture usually persists for some hours, whereafter the muscle will contract. When it becomes hard and stiff the whole body becomes inflexible and the fish is in rigor mortis This condition usually lasts for a day or more and then rigor resolves. The resolution of rigor mortis makes the muscle relax again and it becomes limp, but no longer as elastic as before rigor. The rate in onset and resolution of rigor varies from species to species and is affected by temperature, handling, size and physical condition of the fish. The effect of temperature on rigor is not uniform. In the case of cod, high temperatures give a fast onset and a very strong rigor mortis. This should be avoided as strong rigor tensions may cause gaping, *i.e.*, weakening of the connective tissue and rupture of the fillet.

Table: Onset and Duration of Rigor Mortis in Various Ffish Species

Species	Condition	Temperature °C	Time from death to onset of *rigor* (hours)	Time from death to end of *rigor* (hours)
Cod *(Gadus morhua)*	Stressed	0	2-8	20-65
	Stressed	10-12	1	20-30
	Stressed	30	0.5	1-2
	Unstressed	0	14-15	72-96
Grouper *(Epinephelus malabaricus)*	Unstressed	2	2	18
Blue Tilapia *(Areochromis aureus)*	Stressed	0	1	
	Unstressed	0	6	
Tilapia *(Tilapia mossanibica)* small 60g	Unstressed	0-2	2-9	26.5
Grenadier *(Macrourus whitson)*	Stressed	0	<1	35-55
Anchovy *(Engraulis anchoita)*	Stressed	0	20-30	18
Plaice *(Pleuronectes platessa)*	Stressed	0	7-11	54-55
Coalfish *(Pollachius virens)*	Stressed	0	18	110
Redfish *(Sebastes spp.)*	Stressed	0	22	120
Japanese flounder *(Paralichthys olivaceus)*		0	3	>72
		5	12	>72
		10	6	72
		15	6	48
		20	6	24
Carp *(Cyprinus carpio)*		0	8	
		10	60	
		20	16	
	Stressed	0	1	
	Unstressed	0	6	

It has generally been accepted that the onset and duration of rigor mortis are more rapid at high temperatures, but observations, especially on tropical fish show the opposite effect of temperature with regard to the onset of rigor. It is evident that in these species the onset of rigor is accelerated at 0°C compared to 10°C, which is in good correlation with a stimulation of biochemical changes at 0°C. However, an explanation for this has been suggested by Abe and Okuma (1991) who have shown that onset of rigor mortis in carp (Cyprinus *carpio)* depends on the difference in sea temperature and storage temperature. When the difference is large the time from death to onset of rigor is short and *vice versa.*

Rigor mortis starts immediately or shortly after death if the fish is starved and the glycogen reserves are depleted, or if the fish is stressed. The method used for stunning and killing the fish also influences the onset of rigor. Stunning and killing by hypothermia (the fish is killed in iced water) give the fastest onset of rigor, while a blow on the head gives a delay of up to 18 hours.

SENSORY EVALUATION OF RAW FISH IN MARKETS

The technological significance of rigor mortis is of major importance when the fish is filleted before or in rigor. In rigor the fish body will be completely stiff; the filleting yield will be very poor, and rough handling can cause gaping. If the fillets are removed from the bone pre-rigor the muscle can contract freely and the fillets will shorten following the onset of rigor. Dark muscle may shrink up to 52 % and white muscle up to 15 % of the original length . If the fish is cooked pre-rigor the texture will be very soft and pasty.

In contrast, the texture is tough but not dry when the fish is cooked in rigor. Post-rigor the flesh will become firm, succulent and elastic. Whole fish and fillets frozen pre-rigor can give good products if they are carefully thawed at a low temperature in order to give *rigor mortis* time to pass while the muscle is still frozen. The sensory evaluation of raw fish in markets and landing sites is done by assessing the appearance, texture and odour. Most scoring systems are based upon changes taking place during storage in melting ice. It should be remembered that the characteristic changes vary depending on the storage method.

The appearance of fish stored under chilled condition without ice does not change as much as for iced fish, but the fish spoil more rapidly and an evaluation of cooked flavour will be necessary. A knowledge of the time/temperature history of the fish should therefore be essential at landing. The characteristic sensory changes in fish post mortem vary considerably depending on fish species and storage method.

A general description has been provided by the EEC in the guidelines for quality assessment of fish. The suggested scale is numbered from 0 to 3, where 3 is the best quality.

The West European Fish Technologists' Association has compiled a multilingual glossary of odours and flavours which also can be very useful when looking for descriptive words for sensory evaluation of freshness of fish.

CHANGES IN EATING QUALITY

If quality criteria of chilled fish during storing are needed, sensory assessment of the cooked fish can be conducted. A characteristic pattern of the deterioration of fish stored in ice can be found and divided into the following four phases:

Table: Freshness ratings: Council Regulation

Criteria				
	Marks			
Part of fish inspected	**3**	**2**	**1**	**0**
Appearance				
Skin	Bright, iridescent pigmentation, no discoloration Aqueous, transparent, mucus	Pigmentation bright but not lustrous Slightly cloudy mucus	Pigmentation in the process of becoming discoloured and dull Milky mucus	[1]Dull pigmentation Opaque mucus
Eye	Convex (bulging) Transparent cornea Black, bright pupil	Convex and slightly sunken Slightly opalescent cornea Black, dull pupil	Flat Opalescent cornea Opaque pupil	[1]Concave in the centre Milky cornea Grey pupil
Gills	Bright colour No mucus	Less coloured Slight traces of clear mucus	Becoming discoloured Opaque mucus	[1]Yellowish Milky mucus
Flesh (cut from abdomen)	Bluish, translucent, smooth, shining No change in original colour	Velvety, waxy, dull Colour slightly changed	Slightly opaque	[1]Opaque
Colour (along vertebral column)	Uncoloured	Slightly pink	Pink	[1]Red
Organs	Kidneys and residues of other organs should be bright red, as should the blood inside the aorta	Kidneys and residues of other organs should be dull red; blood becoming discoloured	Kidneys and residues of other organs and blood should be pale red	Kidneys and residues of other organs and should be brownish in colour
Condition				
Flesh	Firm and elastic Smooth surface	Less elastic	Slightly soft (flaccid), less elastic Waxy (velvety) and dull surface	[1]Soft (flaccid) Scales easily detached from skin, surface rather wrinkled, inclining to mealy
Vertebral column	Breaks instead of coming away	Sticks	Sticks slightly	[1]Does not stick
Peritoneum	Sticks completely	Sticks	Sticks slightly	[1]Does not stick

[1] Or in a more advanced state of decay.

- Phase 1 The fish is very fresh and has a sweet, seaweedy and delicate taste. The taste can be very slightly metallic. In cod, haddock, whiting and flounder, the sweet taste is maximized 2-3 days after catching.
- Phase 2 There is a loss of the characteristic odour and taste. The flesh becomes neutral but has no off-flavours. The texture is still pleasant.
- Phase 3 There is sign of spoilage and a range of volatile, unpleasant-smelling substances is produced depending on the fish species and type of spoilage (aerobic, anaerobic). One of the volatile compounds may be trimethylamine (TMA) derived from the bacterial reduction of trimethyl-aminoxide (TMAO). TMA has a very characteristic "fishy" smell. At the beginning of the phase the off-flavour may be slightly sour, fruity and slightly bitter, especially in fatty fish. During the later stages sickly sweet, cabbage-like, ammoniacal, sulphurous and rancid smells develop. The texture becomes either soft and watery or tough and dry.
- Phase 4 The fish can be characterized as spoiled and putrid.

A numbered scale may be used for the sensory evaluation of cooked fish. The scale is numbered from 0 to 10, 10 indicating absolute freshness, 8 good quality and 6 a neutral tasteless fish. The rejection level is 4. Using the scale in this way the graph becomes S-shaped indicating a fast degradation of the fish during the first phase, a slower rate in phase 2 and 3 and finally a high rate when the fish is spoiled. Other scales can well be used and can change the shape of the graph. It is, however, important to understand the kind of results desired from the sensory analysis in order to ask the right questions to the sensory assessors.

Autolytic Changes

Autolysis means "self-digestion". It has been known for many years that there are at least two types of fish spoilage: bacterial and enzymatic. Uchyama and Ehira (1974) showed that for cod and yellowtail tuna, enzymatic changes related to fish freshness preceded and were unrelated to changes in the microbiological quality. In some species (squid, herring), the enzymatic changes precede and therefore predominate the spoilage of chilled fish. In others, autolysis contributes to varying degrees to the overall quality loss in addition to microbially-mediated processes.

PRODUCTION OF ENERGY IN POST MORTEM MUSCLE

At the point of death, the supply of oxygen to the muscle tissue is interrupted because the blood is no longer pumped by the heart and is not circulated through the gills where, in the living fish, it becomes enriched with oxygen. Since no oxygen is available for normal respiration, the production of energy from ingested nutrients is greatly restricted. Glycogen (stored carbohydrate) or fat is oxidized or "burned" by the tissue enzymes in a series of reactions which ultimately produce carbon dioxide (CO_2), water and the energy-rich organic compound adenosine triphosphate (ATP). This type of respiration takes place in two stages: an anaerobic and an aerobic stage. The latter depends on the continued presence of oxygen (O_2) which is only available from the circulatory system. Most crustaceans are capable of respiring outside the aquatic environment by absorption of atmospheric oxygen for limited periods.

ATP may be synthesized by two other important pathways from creatine phosphate or from arginine phosphate. The former source of energy is restricted to vertebrate muscle (teleost fish) while the latter is characteristic of some invertebrates such as the cephalopods (squid and octopus). In either case, ATP production ceases when the creatine or arginine phosphates are depleted. It is interesting to note that octopine is the end-product from the anaerobic metabolism of cephalopods and is not acidic (unlike lactate), thus any changes in post mortem pH in such animals are not related to the production of lactic acid from glycogen.

For most teleost fish, glycolysis is the only possible pathway for the production of energy once the heart stops beating. This more inefficient process has principally lactic and pyruvic acids as its end-products. In addition, ATP is produced in glycolysis, but only 2 moles for each mole of glucose oxidized as compared to 36 moles ATP produced for each mole of glucose if the glycolytic end products are oxidized aerobically in the mitochondrion in the living animal. Thus, after death, the anaerobic muscle cannot maintain its normal level of ATP, and when the intracellular level declines from 7-10 μmoles/g to £ 1.0 μmoles/g tissue, the muscle enters rigor mortis. Post mortem glycolysis results in the accumulation of lactic acid which in turn lowers the pH of the muscle. In cod, the pH drops from 6.8 to an ultimate pH of 6.1-6.5. In some species of fish, the final pH may be lower: in large mackerel, the ultimate rigor pH may be as low as 5.8-6.0 and as low as 5.4-5.6 in tuna and halibut, however such low pH levels are unusual in marine teleosts.

These pHs are seldom as low as those observed for post mortem mammalian muscle. For example, beef muscle often drops to pH levels of 5.1 in rigor mortis. The amount of lactic acid produced is related to the amount of stored carbohydrate (glycogen) in the living tissue. In general, fish muscle contains a relatively low level of glycogen compared to mammals, thus far less lactic acid is generated after death.

Also, the nutritional status of the fish and the amount of stress and exercise encountered before death will have a dramatic effect on the levels of stored glycogen and consequently on the ultimate post mortem pH. As a rule, well-rested, well-fed fish contain more glycogen than exhausted fish. In a recent study of Japanese loach, it was shown that only minutes of pre-capture stress resulted in a decrease of 0.50 pH units in 3 hours as compared to non-struggling fish whose pH dropped only 0.10 units in the same time period. In addition, the same authors showed that bleeding of fish significantly reduced the post mortem production of lactic acid.

The post mortem reduction in the pH of fish muscle has an effect on the physical properties of the muscle. As the pH drops, the net surface charge on the muscle proteins is reduced, causing them to partially denature and lose some of their water-holding capacity. Muscle tissue in the state of rigor mortis loses its moisture when cooked and is particularly unsuitable for further processing which involves heating, since heat denaturation enhances the water loss. Loss of water has a detrimental effect on the texture of fish muscle and it has been shown by Love (1975) that there is an inverse relationship between muscle toughness and pH, unacceptable levels of toughness (and water-loss on cooking) occurring at lower pH levels.

Autolysis and Nucleotide Catabolism

The rigor mortis sets in when the muscle ATP level drops to £ 1.0 μmoles/ g. ATP is not only a source of high energy which is required for muscle contraction in the living animal, but also acts as a muscle plasticizer. Muscle contraction per se is controlled by calcium and an enzyme, ATP-ase which is found in every muscle cell. When intracellular Ca^{+2} levels are 1 μM, Ca^{+2} - activated ATP-ase reduces the amount of free muscle ATP which results in the interaction between the major contractile proteins, actin and myosin. This ultimately results in the shortening of the muscle, making it stiff and inextensible. A fish in rigor mortis cannot normally be filleted or processed because the carcass is too stiff to be manipulated and is often contorted, making machine-handling impossible.

The resolution of rigor is a process still not completely understood but always results in the subsequent softening (relaxation) of the muscle tissue and is thought to be related to the activation of one or more of the naturally-occurring muscle enzymes, digesting away certain components of the rigor mortis complex. The softening of the muscle during resolution of rigor (and eventually spoilage processes) is coincidental with the autolytic changes. Among the changes, one of the first to be recognized was the degradation of ATP-related compounds in a more-or-less predictable manner after death.

The degradation of ATP catabolites proceeds in the same manner with most fish but the speed of each individual reaction (from one catabolite to another) greatly varies from one species to another and often progresses coincidentally with the perceived level of spoilage as determined by trained analysts. Saito *et al.* (1959) were the first to observe this pattern and to develop a formula for fish freshness based on these autolytic changes:

$$K\% = \frac{[Ino]+[Hx]}{[ATP]+[ADP]+[AMP]+[IMP]+[Ino]+[Hx]} \times 100$$

where [ATP], [ADP], [AMP], [IMP], [Ino] and [Hx] represent the relative concentrations of these compounds in fish muscle measured at various times during chilled storage. The K or "freshness" index gives a relative freshness rating based primarily on the autolytic changes which take place during post mortem storage of the muscle. Thus, the higher the K value, the lower the freshness level. Unfortunately, some fish species such as Atlantic cod reach a maximum K value well in advance of the shelf life as determined by trained judges, and K is therefore not considered reliable as a freshness index for all marine finfish.

Also, the degradation of nucleotide catabolites is only coincidental with perceived changes in freshness and not necessarily related to the cause of freshness deterioration since only Hx is considered to have a direct effect on the perceived bitter off-flavour of spoiled fish . It is now widely accepted that IMP is responsible for the desirable fresh fish flavour which is only present in top quality seafood. None of the nucleotide catabolites are considered to be related to the perceived changes in texture during the autolytic process except of course ATP whose loss is associated with *rigor mortis.* Surette *et al.* (1988) followed the autolysis of sterile and non-sterile cod as indicated by the ATP catabolites. The rates of formation and breakdown of IMP were the same in both sterile and non- sterile samples of cod tissue, indicating that the catabolic pathway for the degradation of ATP through to inosine is entirely due to autolytic enzymes.

The conversion of Iino to Hex was accelerated by about 2 days for the non-sterile samples, suggesting that bacterial nucleoside phosphorylase plays a major role in the *postmortem* production of Hx in refrigerated cod. It is interesting to note that Surette *et al.* (1988) were not able to recover nucleoside phosphorylase from freshly killed cod, but Surette *et al.* (1990) later went on to isolate and purify this enzyme from a *Proteus* bacterium recovered from spoiled cod fillets. As mentioned earlier, large variations can be expected in the patterns of nucleotide degradation from one species to another.

There is little doubt that physical handling accelerates the autolytic changes in chilled fish. Surette *et al.* (1988) reported that the breakdown rate of the nucleotide catabolites was greater in sterile fillets than in non-sterile gutted whole cod.

This is perhaps not surprising since many of the autolytic enzymes have been shown to be compartmentalized in discrete membrane-bound packages which become broken when subjected to physical abuse and result in the intimate mixing of enzyme and substrate. Crushing of the fish by ice or other fish can seriously affect the edibility and filleting yields even for fish which have a relatively low bacterial load, demonstrating the importance of autolytic processes. Iced fish should never be stored in boxes deeper than 30 cm and it is equally important to be sure that fish boxes are not permitted to "nest" one on top of the other if autolysis is to be minimized. Systems for conveying fish and for discharge from the vessels must be designed so as to avoid physical damage to the delicate tissues.

Several rapid methods have been developed for the determination of individual nucleotide catabolites or combinations including the freshness index. Two recent reviews should be consulted.

AUTOLYTIC CHANGES INVOLVING PROTEOLYTIC ENZYMES

Many proteases have been isolated from fish muscle and the effects of proteolytic breakdown are often related to extensive softening of the tissue. Perhaps one of the most notable examples of autolytic proteolysis is the incidence of belly-bursting in pelagic (fatty fish) species such as herring and capelin. This type of tissue softening is most predominant in summer months when pelagics are feeding heavily, particularly on "red feed" consisting of copepods and euphausiids. The low molecular weight peptides and free amino-acids produced by the autolysis of proteins not only lower the commercial acceptability of pelagics, but in bulk-stored capelin, autolysis has been shown to accelerate the growth of spoilage bacteria by providing a superior growth environment for such organisms . The induction of bacterial spoilage in capelin by autolysis also resulted in the decarboxylation of amino-acids, producing biogenic amines and lowered the nutritive value of the fish significantly. This is particularly important since autolysis and bacterial growth greatly lower the commercial value of pelagics used for the production of fishmeal.

Similarly, bulk-stored herring used for fishmeal has been found to contain carboxy-peptidases A and B, chymotrypsin, and trypsin; and preliminary studies have shown that proteolysis can be inhibited by the addition of potato extracts which not only slowed the proteolysis but resulted in lower microbial growth and preservation of the nutritional value of the meal .

More recently, Botta *et al.* (1992) found that autolysis of the visceral cavity (belly-bursting) of herring was related more to physical handling practices than to biological factors such as fish size, amount of red feed in the gut or roe content. In particular, it was shown that for herring, freezing/thawing, thawing time at 15°C and time of iced storage, had a far greater influence on belly-bursting than biological factors.

Several Proteolytic Enzymes

Although several proteolytic enzymes have been discovered in the fish tissues, it has perhaps been the cathepsins which have been described most often. The cathepsins are "acid" proteases usually found packaged in tiny, submicroscopic organelles called lysozomes. In living tissue, lysozomal proteases are believed to be responsible for protein breakdown at sites of injury. Thus cathepsins are for the most part inactive in living tissue but become released into the cell juices upon physical abuse or upon freezing and thawing of *post mortem* muscle.

Cathepsins D and L are believed to play a major role in the autolytic degradation of fish tissue since most of the other cathepsins have a relatively narrow pH range of activity far too low to be of physiological significance. Reddi *et al.* (1972) demonstrated that an enzyme believed to be cathepsin D from winter flounder was active over a pH range of 3-8 with a maximum near pH 4.0, although no attempt was made to confirm the identity of the enzyme using synthetic substrates or specific inhibitors. Nevertheless, the enzyme was far less active in the presence of ATP, suggesting that such an enzyme would only be active in *post mortem* fish muscle. Also, the enzyme activity was inhibited strongly by the presence of salt with virtually no activity remaining after a 25-hour incubation in the presence of 5% sodium chloride. It is therefore unlikely that Reddi's enzyme was active in salted fish products.

Cathepsin L has been implicated in the softening of salmon muscle during spawning migration. It is likely that this enzyme contributes more to autolysis of fish muscle than cathepsin D since it is far more active at neutral pH, and has been shown to digest both myofibrillar proteins (actomyosin) as well as connective tissue. Yamashita and Konogaya (1990) produced strong evidence implicating cathepsin L rather than other cathepsins in the softening of salmon during spawning.

They demonstrated that electrophoresis of purified myofibrils treated with cathepsin L resulted in patterns which were almost identical to patterns of proteins recovered from muscle from spawning fish. Furthermore, the cathepsin L autolytic activity correlated well with the texture of the muscle as measured instrumentally. The linear correlation between cathepsin L activity and breaking strength of the muscle was excellent; r = 0.86 and -0.95 for fresh and frozen/thawed tissue, respectively.

It is interesting that, in all cases, the autolytic ability as measured by cathepsin L activity was higher in frozen/thawed tissue than in fresh tissue. Freezing and thawing often break down cell membranes allowing autolytic membrane-bound enzymes to react with their natural substrates. The enzyme and its naturally occurring inhibitor were further studied by the same authors. Cathepsin L has also been associated with the production of a jelly- like softening of flounder and the uncontrollable softening of Pacific hake muscle which has been parasitized by Myxosporidia . The tissues of such infected fish have little commercial value, but at present it is not known if it is the parasite or the host which secretes the proteolytic enzymes which autolyze the muscle. In addition to their detrimental effect on texture, catheptic enzymes induce intentional autolytic changes in fermented fish products. For example, cathepsins are believed to be responsible for major textural changes during the fermentation of salted preserved Japanese squid and Crucian carp.

Intracellular Proteases

A second group of intracellular proteases called "calpains" or "calcium activated factor" (CAF) has recently been associated with fish muscle autolysis and is found in meats, finfish and crustaceans. Tenderness is probably the most important quality characteristic of red meat. It has been known for nearly a century that *post mortem* aging of red meat results in the tenderization process. Calpains have been found primarily responsible for the *post mortem* autolysis of meat through digestion of the z- line proteins of the myofibril. Although toughness is seldom a problem with unfrozen fish muscle, softening through autolysis is a serious problem limiting the commercial value. The calpains are intra-cellular endopeptidases requiring cysteine and calcium; μ-calpain requiring 5-50 μM Ca^{2+}, m-calpain requiring 150-1000 μM Ca^{2+}. Most calpains are active at physiological pH, making it reasonable to suspect their importance in fish-softening during chilled storage. Studies have shown that in crustacean muscle, calpains are associated with moltinduced textural changes to the muscle and carry out non-specific generalized digestion of the myofibrillar proteins.

However, vertebrate muscle calpains have been shown to be very specific, digesting primarily tropinin- T, desmin, titin and nebulin, attacking neither vertebrate actin or myosin . In contrast, fish calpains digest myosin (specifically the myosin heavy chain) to form an initial fragment with approximate molecular weight of 150 000 Da . The same authors demonstrated that fish calpains were far more active at low temperatures than were mammalian calpains and that the rates of cleavage were species-specific, being most active against myosins with lowest heat stabilities. Thus, fish species adapted to colder environmental temperatures are more susceptible to calpain autolysis than those from tropical waters. Although calpain has been identified in several fish species including carp, tilapia and shrimp, as well as tuna, croaker, red seabream and trout to name a few, little work has to date demonstrated a "cause and effect" relationship between calpain activity and instrumental measurements of texture.

Collagenases

To this point, all of the *post mortem* autolytic changes described have involved changes within the muscle cell *per se.* However, the flesh of teleost fish is divided into blocks of muscle cells separated into "flakes" or myotomes by connective tissue called myocommata. Each muscle cell or fibre is surrounded with connective tissue which attaches to the myocommata at the ends of the cells by means of fine collagenous fibrils. During chilled storage, these fibrils deteriorate . More recently, it was shown that instrumental measurements of texture of chilled trout muscle decreased as the amount of type V collagen was solubilized, presumably due to the action of autolytic collagenase enzymes . It is these enzymes which presumably cause "gaping" or breakdown of the myotome during long-term storage on ice or short term storage at high temperature. For Atlantic cod, it has been shown that upon reaching 17°C, gaping is inevitable presumably because of degradation of the connective tissue and rapid shortening of the muscle due to high temperature *rigor.* The relatively short shelf life of chilled prawns due to softening of the tissue has also been shown to be due to the presence of collagenase enzymes . The source of the collagenase enzymes in prawn is thought to be the hepatopancreas (digestive organ).

AUTOLYTIC CHANGES DURING FROZEN STORAGE

The reduction of trimethylamine oxide (TMAO), an osmoregulatory compound in many marine teleost fish, is usually due to bacterial action but in some species an enzyme is present in the muscle tissue which is able to break down TMAO into dimethylamine (DMA) and formaldehyde (FA):

$\rightarrow {}_2NH + HCHO$

It is important to note that the amount of formaldehyde produced is equivalent to the dimethylamine formed but is of far greater commercial significance. Formaldehyde induces cross- linking of the muscle proteins making the muscle tough and readily lose its water holding capacity. The enzyme responsible for formal dehyde-induced toughening is called TMAO-ase or TMAO demethylase and is most commonly found in the gadoid fishes (cod family). Most of the TMAO demethylase enzymes reported to date weremembrane-bound and become most active when the tissue membranes are disrupted by freezing or artificially by detergent solubilization.

Dark (red) muscle has a higher rate of activity than white muscle whereas other tissues such as kidney, spleen and gall bladder are extremely rich in the enzyme. Thus, it is important that minced fish is completely free of organ tissue such as kidney from gadoid species if toughening in frozen storage is to be avoided.

It is often difficult to ensure that the kidney is removed prior to mechanical deboning since this particular organ runs the full length of the backbone and is adherent to it. The TMAO-ase enzyme has been isolated from the microsomal fraction in hake muscle and the lysosomal membrane in kidney tissue . It has been shown that the toughening of frozen hake muscle is correlated to the amount of formaldehyde produced, and that the rate of FA production is greatest at high frozen-storage temperatures .

In addition, it has been shown that the amount of FA-induced toughening is enhanced by physical abuse to the catch prior to freezing and by temperature fluctuations during frozen storage. The most practical means of preventing the autolytic production of FA is to store fish at temperatures $< -30°C$ to minimize temperature fluctuations in the cold store and to avoid rough handling or the application of physical pressure on the fish prior to freezing. Generally, the most important single factor affecting autolysis is physical disruption of the muscle cells. No attempt has been made here to deal with the alkaline proteases associated with the softening of cooked surimi products. An article by Kinoshita *et al.* (1990) deals with the heat-activated alkaline proteases associated with the softening in surimi-based products.

FISH LIPIDS OF IMPORTANCE

They result in production of a range of substances among which some have unpleasant (rancid) taste and smell. Some may also contribute to texture changes by binding covalently to fish muscle proteins. The various reactions are either *nonenzymatic* or catalyzed by *microbial* enzymes or by *intracellular* or *digestive* enzymes from the fish themselves. The relative significance of these reactions, therefore, mainly depends on fish species and storage temperature. Fatty fish are, of course, particularly susceptible to lipid degradation which can create severe quality problems even on storage at subzero temperatures.

OXIDATION

The large amount of polyunsaturated fatty acid moieties found in fish lipids makes them highly susceptible to oxidation by an autocatalytic mechanism. The process is initiated as described below by abstraction of a hydrogen atom from the central carbon of the *pentadiene structure* found in most fatty acid acyl chains containing more than one double bond:

$$-CH = CH-CH_2-CH = CH- \rightarrow -CH = CH-CH-CH = CH- + H$$

Contrary to the native molecule, the lipid radical (L) reacts very quickly with atmospheric oxygen making a peroxy-radical (LOO) which again may abstract a hydrogen from another acyl chain resulting in a lipid hydroperoxide (LOOH) and a new radical L.

This propagation continues until one of the radicals is removed by reaction with another radical or with an *antioxidant* (AH) whose resulting radical (A) is much less reactive. The hydroperoxides produced in relatively large amounts during propagation are tasteless, and it is therefore perhaps not surprising that the widely used "peroxide value" usually correlates rather poorly to sensorial properties.

The hydroperoxides are readily broken down, catalyzed by heavy metal ions, to secondary autoxidation products of shorter carbon chain-length. These secondary products - mostly aldehydes, ketones, alcohols, small carboxylic acids and alkanes - give rise to a very broad odour spectrum and in some cases to a yellowish discoloration. Several of the aldehydes can be determined as "thiobarbituric acid-reactive substances".

Metal ions are very important in the first step of lipid autoxidation - the initiation process - in catalyzing the formation of reactive oxygen species as for example the hydroxyl radical (OH). This radical immediately reacts with lipids or other molecules at the site where it is generated. The high reactivity may explain that free fatty acids have been found to be more susceptible to oxidation than the corresponding bound ones, because the amount of iron in the aqueous phase is probably greater than the amount bound to the surface of cellular membranes and lipid droplets. Fatty acid hydroperoxides may also be formed enzymatically, catalyzed by *lipoxygenase* which is present in variable amounts in different fish tissues. A relatively high activity has been found in the gills and under the skin of many species. The enzyme is unstable and is probably important for lipid oxidation only in fresh fish. Cooking or freezing/ thawing rather effectively destroys the enzyme activity.

The living cells possess several protection mechanisms directed against lipid oxidation products. An enzyme, glutathione peroxidase, exists which reduces hydroperoxides in the cellular membranes to the corresponding hydroxy-compounds.

This reaction demands supply of reduced glutathione and will therefore cease post mortem when the cell is depleted of that substance. The membranes also contain the phenolic compound a-tocopherol (Vitamin E) which is considered the most important natural antioxidant. Tocopherol can donate a hydrogen atom to the radicals L- or LOO- functioning as the molecule AH. It is generally assumed, that the resulting tocopheryl radical reacts with ascorbic acid (Vitamin C) at the lipid/water interface regenerating the tocopherol molecule. Other compounds, for example the carotenoids, may also function as antioxidants. Wood smoke contains phenols which may penetrate the fish surface during smoking and thereby provide some protection against lipid oxidation.

HYDROLYSIS

During storage, a considerable amount of free fatty acids (FFA) appears. The phenomenon is more profound in ungutted than in gutted fish probably because of the involvement of digestive enzymes. Triglyceride in the depot fat is cleaved by triglyceride lipase originating from the digestive tract or excreted by certain microorganisms. Cellular lipases may also play a minor role.

In lean fish, for example Atlantic cod, production of free fatty acids also occurs, even at low temperatures. The enzymes responsible are believed to be cellular phospholipases - in particular phospholipase A_2 - although a correlation between activity of these enzymes and the rate of appearance of FFA has as yet not been firmly established. The fatty acids bound to phospholipids at glycerol-carbon atom 2 are largely of the polyunsaturated type, and hydrolysis therefore often leads to increased oxidation as well. Furthermore, the fatty acids themselves may cause a "soapy" off-flavour.

CHEMICAL COMPOSITION AND PRINCIPAL CONSTITUENTS

The chemical composition of fish varies greatly from one species and one individual to another depending on age, sex, environment and season.

The principal constituents of fish and mammals may be divided into the same categories, and examples of the variation between the constituents in fish are shown in Table below. The composition of beef muscle has been included for comparison.

Table: Principal Constituents (percentage) of Fish and Beef Muscle

Constituent	Fish (fillet)			Beef (isolated muscle)
	Min.	Normal variation	Max.	
Protein	6	16-21	28	20
Lipid	0.1	0.2-25	67	3
carboydrate		<0.5		1
Ash	0.4	1.2-1.5	105	1
Water	28	66-81	96	75

SOURCES: Stansby, 1962; Love, 1970

As can be seen from Table, a substantial normal variation is observed for the constituents of fish muscle. The minimum and maximum values listed are rather extreme and encountered more rarely. The variation in the chemical composition of fish is closely related to feed intake, migratory swimming and sexual changes in connection with spawning.

Fish will have starvation periods for natural or physiological reasons (such as migration and spawning) or because of external factors such as shortage of food. Usually spawning, whether occurring after long migrations or not, calls for higher levels of energy. Fish having energy depots in the form of lipids will rely on this. Species performing long migrations before they reach specific spawning grounds or rivers may utilize protein in addition to lipids for energy, thus depleting both the lipid and protein reserves, resulting in a general reduction of the biological condition of the fish. Most species, in addition, do usually not ingest much food during spawning migration and are therefore not able to supply energy through feeding. During periods of heavy feeding, at first the protein content of the muscle tissue will increase to an extent depending upon how much it has been depleted, *e.g.*, in relation to spawning migration. Then the lipid content will show a marked and rapid increase. After spawning the fish resumes feeding behaviour and often migrates to find suitable sources of food. Plankton-eating species such as herring will then naturally experience another seasonal variation than that caused by spawning, since plankton production depends on the season and various physical parameters in the oceans.

The lipid fraction is the component showing the greatest variation. Often, the variation within a certain species will display a characteristic seasonal curve with a minimum around the time of spawning. Although the protein fraction is rather constant in most species, variations have been observed such as protein reduction occuring in salmon during long spawning migrations and in Baltic cod during the spawning season, which for this species extends from January to June/July .

Some tropical fish also show a marked seasonal I variation in chemical composition. West African shad *(Ethmalosa dorsalis)* shows a range in fat content of 2-7 % (wet weight) over the year with a maximum in July . Corvina *(Micropogon furnieri)* and pescada-foguete*(Marodon ancylodon)* captured off the

Brazilian coast had a fat content range of 0.2-8.7 % and 0.1-5.4 % respectively . It has also been observed that the oil content of these species varies with size, larger fish containing about 1 % more oil than smaller ones. Watanabe (1971) examined freshwater fish from Zambia and found a variation from 0.1 to 5.0 % in oil content of four species including both pelagics and demersals. A possible method for discriminating lean from fatty fish species is to term fish that store lipids only in the liver as lean, and fish storing lipids in fat cells distributed in other body tissues as fatty fish.

Typical lean species are the bottom-dwelling ground fish like cod, saithe and hake. Fatty species include the pelagics like herring, mackerel and sprat. Some species store lipids in limited parts of their body tissues only, or in lower quantities than typical fatty species, and are consequently termed semi- fatty species (*e.g.*, barracuda, mullet and shark). The lipid content of fillets from lean fish is low and stable whereas the lipid content in fillets from fatty species varies considerably. However, the variation in the percentage of fat is reflected in the percentage of water, since fat and water normally constitute around 80 % of the fillet. As a rule of thumb, this can be used to estimate the fat content from an analysis of the amount of water in the fillet. In fact, this principle is being utilized with success in a fat-analysing instrument called the Torry Fish Fat Meter, where it is the water content that is actually being measured .

Whether a fish is lean or fatty the actual fat content has consequences for the technological characteristics postmortem. The changes taking place in fresh lean fish may be predicted from knowledge of biochemical reactions in the protein fraction, whereas in fatty species changes in the lipid fractions have to be included. The implication may be that the storage time is reduced due to lipid oxidation, or special precautions have to be taken to avoid this.

The carbohydrate content in fish muscle is very low, usually below 0.5 %. This is typical for striated muscle, where carbohydrate occurs in glycogen and as part of the chemical constituents of nucleotides. The latter is the Source of ribose liberated as a consequence of the autolytic changes *post mortem*.

The chemical composition of the different fish species will show variation depending on seasonal variation, migratory behaviour, sexual maturation, feeding cycles, etc. These factors are observed in wild, free-living fishes in the open sea and inland waters. Fish raised in aquaculture may also show variation in chemical composition, but in this case several factors are controlled, thus the chemical composition may be predicted. To a certain extent the fish farmer is able to design the composition of the fish by selecting the farming conditions. It has been reported that factors such as feed composition, environment, fish size, and genetic traits all have an impact on the composition and quality of the aquacultured fish . The single factors having the most pronounced Impact on the chemical composition is considered to be the feed composition. The fish farmer is interested in making the fish grow as fast as possible on a minimum amount of feed, as the feed is the major cost component in aquaculture. The growth potential is highest when the fish is fed a diet with a high lipid content for energy purposes and a high amount of protein containing a well balanced composition of amino acids.

However, the basic metabolic pattern of the fish sets some limits as to how much lipid can be metabolized relative to protein. Because protein is a much more expensive feed ingredient than lipid, numerous experiments have been performed in order to substitute as much protein as possible with lipids.

Usually most fish species will use some of the protein for energy purposes regardless of the lipid content. When the lipid content exceeds the maximum that can be metabolized for energy purposes, the remainder will be deposited in the tissues, resulting in a fish with very high fat content. Apart from having a negative impact on the overall quality, it may also decrease the yield, as most surplus fat will be stored in depots in the belly cavity, thus being discarded as waste after evisceration and filleting. A normal way of reducing the fat content of aquacultured fish before harvesting is to starve the fish for a period. It has been demonstrated for both fatty and lean fish species that this affects the lipid content.

It should be mentioned that in addition to allowing for the possibility of, within certain limits, predetermining the fish composition in aquaculture operations, keeping fish in captivity under controlled conditions also offers the possibility of conducting experiments in which variation in chemical composition observed in wild fish may be provoked. The experiments may be designed such that the mechanisms behind the variations observed in wild fish may be elucidated.

LIPIDS IN TELEOST FISH SPECIES

The lipids present in teleost fish species may be divided into two major groups: the phospholipids and the triglycerides. The phospholipids make up the integral structure of the unit membranes in the cells; thus, they are often called structural lipids. The triglycerides are lipids used for storage of energy in fat depots, usually within special fat cells surrounded by a phospholipid membrane and a rather weak collagen network. The triglycerides are often termed depot fat. A few fish have wax esters as part of their depot fats. The white muscle of a typical lean fish such as cod contains less than 1 % lipids. Of this, the phospholipids make up about 90 % . The phospholipid fraction in a lean fish muscle consists of about 69 % phosphatidyl-choline, 19 % *phosphatidyl-ethanolamine* and 5 % phosphatidyl-serine. In addition, there are several other phospholipids occurring in minor quantities.

The phospholipids are all contained in membrane structures, including the outer cell membrane, the endoplasmic reticulum and other intracellular tubule systems, as well as membranes of the organelles like mitochondria. In addition to phospholipids, the membranes also contain cholesterol, contributing to the membrane rigidity. In lean fish muscle cholesterol may be found in a quantity of about 6 % of the total lipids. This level is similar to that found in mammalian muscle. As already explained, fish species may be categorized as lean or fatty depending on how they store lipids for energy. Lean fish use the liver as their energy depot, and the fatty species store lipids in fat cells througout the body. The fat cells making up the lipid depots in fatty species are typically located in the subcutaneous tissue, in the belly flap muscle and in the muscles moving the fins and tail. In some species which store extraordinarily high amounts of lipids the fat may also be deposited in the belly cavity. Depending on the amount of polyunsaturated fatty acids, most fish fats are more or less liquid at low temperature.

Finally, fat depots are also typically found spread throughout the muscle structure. The concentration of fat cells appears to be highest close to the myocommata and in the region between the light and dark muscle . The dark muscle contains some triglycerides inside the muscle cells even in lean fish, as this muscle is able to metabolize lipids directly as energy. The corresponding light muscle cells are dependent on glycogen as a source of energy for the anaerobic metabolism. In dark muscle the energy reserves are completely catabolized to CO_2 and water, whereas in light muscle lactic acid is formed. The mobilization of energy is much faster in light muscle than in dark muscle, but the formation of lactic acid creates fatigue, leaving the muscle unable to work for long periods at maximum speed. Thus, the dark muscle is used for continuous swimming activities and the light muscle for quick bursts, such as when the fish is about to catch a prey or to escape a predator. An example of the seasonal variation in fat deposition in mackerel and capelin, where it is seen that the lipid content in the different tissues varies considerably. The lipid stores are typically used for long spawning migrations and when building up gonads . When the lipids are mobilized for these purposes there are questions as to whether the different fatty acids present in the triglyceride are utilized selectively. This is apparently not the case in salmon, but in cod a selective utilization of $C_{22:6}$ has been observed .

The phospholipids may also be mobilized to a certain extent during sustained migrations, although this lipid fraction is considered to be conserved much more than the triglycerides. In elasmobranchs, such as sharks, a significant quantity of the lipid is stored in the liver and may consist of fats like diacyl-alkyl-glyceryl esters or squalene. Some sharks may have liver oils with a minimum of 80 % of the lipid as unsaponifiable substance, mostly in the form of squalene.

Fish lipids differ from mammalian lipids. The main difference is that fish lipids include up to 40% of long-chain fatty acids (14-22 carbon atoms) which are highly unsaturated. Mammalian fat will rarely contain more than two double bonds per fatty acid molecule while the depot fats of fish contain several fatty acids with five or six double bonds . The percentage of polyunsaturated fatty acids with four, five or six double bonds is slightly lower in the polyunsaturated fatty acids of lipids from freshwater fish (approximately 70 %) than in the corresponding lipids from marine fish (approximately 88 %), . However, the composition of the lipids is not completely fixed but can vary with the feed intake and season.

In human nutrition fatty acids such as linoleic and linolenic acid are regarded as essential since they cannot be synthesized by the organism. In marine fish, these fatty acids constitute only around 2 % of the total lipids, which is a small percentage compared with many vegetable oils. However, fish oils contain other polyunsaturated fatty acids which are "essential" to prevent skin diseases in the same way as linoleic and arachidonic acid. As members of the linolenic acid family (first double bond in the third position, w-3 counted from the terminal methyl group), they will also have neurological benefits in growing children.

One of these fatty acids, eicosapentaenoic acid (C20:5 w 3), has recently attracted considerable attention because Danish scientists have found this acid high in the diet of a group of Greenland Eskimos virtually free from arteriosclerosis. Investigations in the United Kingdom and elsewhere have documented that eicosapen-taenoic acid in the blood is an extremely potent antithrombotic factor .

PROTEINS IN FISH MUSCLE TISSUE

The proteins in fish muscle tissue can be divided into the following three groups:

1. Structural proteins (actin, myosin, tropormyosin and actomyosin), which constitute 70-80 % of the total protein content (compared with 40 % in mammals). These proteins are soluble in neutral salt solutions of fairly high ionic strength (^{3}0.5 M).
2. Sarcoplasmic proteins (myoalbumin, globulin and enzymes) which are soluble in neutral salt solutions of low ionic strength (<0.15 M). This fraction constitutes 25-30 % of the protein.
3. Connective tissue proteins (collagen), which constitute approximately 3 % of the -protein in teleostei and about 10 % in elasmobranchii (compared with 17 % in mammals).

The structural proteins make up the contractile apparatus responsible for the muscle movement as explained. The amino-acid composition is approximately the same as for the corresponding proteins in mammaliam muscle, although the physical properties may be slightly different. The isoelectric point (pI) is around pH 4.5-5.5. At the corresponding pH values the proteins have their lowest solublity. The conformational structure of fish proteins is easily changed by changing the physical environment. Treatment with high salt concentrations or heat may lead to denaturation, after which the native protein structure has been irreversibly changed. When the proteins are denatured under controlled conditions their properties may be utilized for technological purposes. A good example is the production of surimi-based products, in which the gel forming ability of the myofibrillar proteins is used.

After salt and stabilizers are added to a washed, minced preparation of muscle proteins, and after a controlled heating and cooling procedure the proteins form a very strong gel .

The majority of the sarcoplasmic proteins are enzymes participating in the cell metabolism, such as the anaerobic energy conversion from glycogen to ATP. If the organelles within the muscle cells are broken, this protein fraction may also contain the metabolic enzymes localized inside the endoplasmatic reticulum, mitochondria and lysosomes. The fact that the composition of the sarcoplasmic protein fraction changes when the organelles are broken was suggested as a method for differentiating fresh from frozen fish, under the assumption that the organelles were intact until freezing.

However, it was later stated that these methods should be used with great caution, as some of the enzymes are liberated from the organelles also during iced storage of fish . The proteins in the sarcoplasmic fraction are excellently suited to distinguishing between different fish species, as all the different species have their characteristic band pattern when separated by the isoelectric focusing method. The method was succesfully introduced by Lundstrom (1980) and has been used by many laboratories and for many fish species. A review of the literature is given by Rehbein (1990).

The chemical and physical properties of collagen proteins are different in tissues such as skin, swim bladder and the myocommata in muscle . In general, collagen fibrils form a delicate network structure with varying complexity in the different connective tissues in a pattern similar to that found in mammals.

However, the collagen in fish is much more thermolabile and contains fewer but more labile cross-links than collagen from warm-blooded vertebrates. The hydroxyprolin content is in general lower in fish than in mammals, although a total variation between 4.7 and 10 % of the collagen has been observed .

Different fish species contain varying amounts of collagen in the body tissues. This has led to a theory that the distribution of collagen may reflect the swimming behaviour of the species . Further, the varying amounts and varying types of collagen in different fishes may also have an influence on the textural properties of fish muscle . Borresen (1976) developed a method for isolation of the collagenous network surrounding each individual muscle cell. The structure and composition of these structures has been further characterized in cod by Almaas (1982).

The role of collagen in fish was reviewed by Sikorsky *et al.* (1984). An excellent, more recent review is given by Bremner (1992), in which the most recent literature of the different types of collagen found in fish is presented.

Fish proteins contain all the essential amino-acids and, like milk, eggs and mammalian meat proteins, have a very high biological value.

Table : Essential amino-acids (percentage) in various proteins

Amino-acid	*Fish*	*Milk*	*Beef*	*Eggs*
Lysine	8.8	8.1	9.3	6.8
Tryptophan	1.0	1.6	1.1	1.9
Histidine	2.0	2.6	3.8	2.2
Phenylalanine	3.9	5.3	4.5	5.4
Leucine	8.4	10.2	8.2	8.4
Isoleucine	6.0	7.2	5.2	7.1
Threonine	4.6	4.4	4.2	5.5
Methionine-cystine	4.0	4.3	2.9	3.3
Valine	6.0	7.6	5.0	8.1

SOURCES: Braekkan, 1976; Moustgard, 1957

Cereal grains are ususally low in lysine and/or the sulphur-containing amino-acids (methionine and cysteine), whereas fish protein is an excellent source of these aminoacids. In diets based mainly on cereals, a supplement of fish can, therefore, raise the biological value significantly.

In addition to the fish proteins already mentioned there is a renewed interest in specific protein fractions that may be recovered from by-products, particularly in the viscera. One such example is the basic protein or protamines found in the milt of the male fish. The molecular weight is usually below 10 000 kD and the pI is higher than 10. This is a result of the extreme amino-acid composition that may show as much as 65 % arginine.

The presence of the basic proteins has long been known, and it is also known that they are not present in all fish species . The best sources are salmonids and herring, whereas ground fish like cod are not found to contain protamines. The extreme basic character of protamines makes them interesting for several reasons. They will adhere to most other proteins less basic. Thus they have the effect of enhancing functional properties of other food proteins. However, there is a problem in removing all lipids present in the milt from the protein preparation, as this results in an off-flavour in the concentrations to be used in foods. Another interesting feature of the basic proteins is their ability to prevent growth of microorganisms. This appears to be the most promising use of these basic proteins in the future.

N-CONTAINING EXTRACTIVES

The N-containing extractives can be defined as the water-soluble, low molecular weight, nitrogen- containing compounds of non-protein nature. This NPN-fraction (non-protein nitrogen) constitutes from 9 to 18 % of the total nitrogen in teleosts. The major components in this fraction are: volatile bases such as ammonia and trimethylamine oxide (TMAO), creatine, free amino-acids, nucleotides and purine bases, and, in the case of cartilaginous fish, urea. An example of the distribution of the different compounds in the NPN-fraction in freshwater and marine fish. It should be noted that the composition varies not only from species to species, but also within the species depending on size, season, muscle sample, etc.

TMAO constitutes a characteristic and important part of the NPN-fraction in marine species and deserves further mention. This component is found in all marine fish species in quantities from 1 to 5 % of the muscle tissue (dry weight) but is virtually absent from freshwater species and from terrestrial organisms. One exception was recently found in a study of Nile perch and tilapia from Lake Victoria, where as much as 150-200 mg TMAO/100 g of fresh fish was found .

Although much work has been conducted on the origin and role of TMAO, there is still much to be clarified. Stroem *et al.* (1979) have shown that TMAO is formed by biosynthesis in certain zooplankton species. These organisms possess an enzyme (TMA mono-oxygenase) which oxidizes TMA to TMAO. TMA is commonly found in marine plants as are many other methylated amines (monomethylamine and dimethylamine). Plankton-eating fish may obtain their TMAO from feeding on these zooplankton (exogenous origin). Belinski (1964) and Agustsson and Stroem (1981) have shown that certain fish species are able to synthesize TMAO from TMA, but this synthesis is regarded as being of minor importance.

The TMA-oxidase system is found in the microsomes of the cells and is dependent on the presence of Nicotinamide ademine denucleotide phosphate (NADPH):

$$(CH_3)_3N + NADPH + H^+ + O_2 \rightarrow (CH_3)_3NO + NADP^+ + H_2O$$

It is puzzling that this mono-oxygenase can be widely found in mammals (where it is thought to function as a detoxifier), while most fish have low or no detectable actitity of this enzyme. Japanese research indicates that there is a TMAO-reducing system present in the dark muscle of certain pelagic fishes. The amount of TMAO in the muscle tissue depends on the species, season, fishing ground, etc. In general, the highest amount is found in elasmobranchs and squid (75-250 mg N/100 g); cod have somewhat less (60-120 mg N/100 g) while flatfish and pelagic fish have the least. An extensive compilation of data is given by Hebard *et al.* (1982). According to Tokunaga (1970), pelagic fish (sardines, tuna, mackerel) have their highest concentration of TMAO in the dark muscle while demersal, white-fleshed fish have a much higher content in the white muscle. In elasmobranchs, TMAO seems to play a role in osmoregulation, and it has been shown that a transfer of small rays to a mixture of fresh and sea water (1: 1) will result in a 50 % reduction of intracellular TMAO. The role of TMAO in teleosts is more uncertain. Several hypotheses for the role of TMAO have been proposed:

- TMAO has no significant function. It is accumulated in the muscle when the fish is fed a TMAO-containing diet
- TMAO is essentially a waste product, a detoxified form of TMA

- TMAO is an osmoregulator
- TMAO functions as an "anti-freeze"

According to Stroem (1984), it is now generally believed that TMAO has an osmoregulatory role. As the occurrence of TMAO had previously been found virtually only in marine species until the observation published by Gram *et al.* (1989), it was speculated that TMAO together with high amounts of taurine could have additional effects, at least in fresh water fish .

Quantitatively, the main component of the NPN-fraction is creatine. In resting fish, most of the creatine is phosphorylated and supplies energy for muscular contraction.

The NPN-fraction also contains a fair amount of free amino-acids. These constitute 630 mg/ 100 g light muscle in mackerel *(Scomber scombrus),* 350-420 mg/ 100 g in herring *(Clupea harengus)* and 310-370 mg/100 g in capelin *(Mallotus villosus).* The relative importance of the different amino- acids varies with species. Taurine, alanine, glycine and imidazole-containing amino-acids seem to dominate in most fish. Of the imidazole-containing amino-acids, histidine has attracted much attention because it can be decarboxylated microbiologically to histamine. Active, dark-fleshed species such as tuna and mackerel have a high content of histidine.

AMOUNT OF VITAMINS AND MINERALS

The amount of vitamins and minerals is species-specific and can furthermore vary with season. In general, fish meat is a good source of the B vitamins and, in the case of fatty species, also of the A and D vitamins. Some freshwater species such as carp have high thiaminase actitity so the thiamine content in these species is usually low. As for minerals, fish meat is regarded as a valuable source of calcium and phosphorus in particular but also of iron, copper and selenium. Saltwater fish have a high content of iodine.

The vitamin content is comparable to that of mammals except in the case of the A and D vitamins which are found in large amounts in the meat of fatty species and in abundance in the liver of species such as cod and halibut. It should be noted that the sodium content of fish meat is relatively low which makes it suitable for low-sodium diets.

Table: Some Mineral Constituents of Fish Muscle

Element	*Average value (mg/100 g)*	*Range (mg/100 g)*
Sodium	72	30 -134
Potassium	278	19 -502
Calcium	79	19 -881
Magnesium	38	4.5-452
Phosphorus	190	68-550

SOURCE: Murray and Burt, 1969

In aquacultured fish, the contents of vitamins and minerals are considered to reflect the composition of the corresponding components in the fish feed, although the observed data should be interpreted with great caution . In order to protect the n-3 polyunsaturated fatty acids, considered of great importance both for fish and human health, vitamin E may be added to the fish feed as an antioxidant. It has been shown that the resulting level of vitamin E in the fish tissue corresponds to the concentration in the feed .

BIOLOGICAL ASPECTS AND CLASSIFICATION OF FISH

Fish are generally defined as aquatic vertebrates that use gills to obtain oxygen from water and have fins with variable number of skeletal elements called fin rays . Five vertebrate classes have species which could be called fish, but only two of these groups - the sharks and rays, and the bonyfish - are generally important and widely distributed in the aquatic environment. Fish are the most numerous of the vertebrates, with at least 20000 known species, and more than half (58 %) are found in the marine environment. They are most common in the warm and temperate waters of the continental shelves (some 8 000 species). In the cold polar waters about 1 100 species are found. In the oceanic pelagic environment well away from the effect of land, there are only some 225 species. Surprisingly, in the deeper mesopelagic zone of the pelagic environment (between 100 and 1 000 m depth) the number of species increases. There are some 1 000 species of so-called mid- water fish .

Classifying all these organisms into a system is not an easy task, but the taxonomist groups organisms into natural units that reflect evolutionary relationships. The smallest unit is the species. Each species is identified by a scientific name which has two parts the genus and the specific epithet (binominal nomenclature). The genus name is always capitalized and both are italicized. As an example, the scientific (species) name of the common dolphin is *Delphinus delphis.* The genus is a category that contains one or more species, while the next step in the hierarchy is the family which may contain one or more genus. Thus the total hierarchical system is: Kingdom: Phylum: Class: Order: Family: Genus: Species.

The use of common or local names often creates confusion since the same species may have different names in different regions or, conversely, the same name is ascribed to several different species, sometimes with different technological properties. As a point of reference the scientific name should, therefore, be given in any kind of publication or report the first time a particular species is referred to by its common name. For further information see the International Council for the Exploration of the Sea "List of names of Fish and Shellfish" ; the "Multilingual Dictionary of Fish and Fish Products" prepared by the Organisation for Economic Cooperation and Development and the "Multilingual Illustrated Dictionary of Aquatic Animals and Plants" (Commission of the European Communities, 1993). The classification of fish into cartilaginous and bony (the jawless fish are of minor importance) is important from a practical viewpoint, since these groups of fish spoil differently and vary with regard to chemical composition.

ANATOMY AND PHYSIOLOGY

The Skeleton

Being vertebrates, fish have a vertebral column - the backbone - and a cranium covering the brain. The backbone runs from the head to the tail fin and is composed of segments (vertebrae). These vertebrae are extended dorsally to form neural spines, and in the trunk region they have lateral processes that bear ribs. The ribs are cartilaginous or bony structures in the connective tissue (myocommata) between the muscle segments (myotomes). Usually, there is also a corresponding number of false ribs or "pin bones" extending more or less horizontally into the muscle tissue. These bones cause a great deal of trouble when fish are being filleted or otherwise prepared for food.

Muscle Anatomy and Function

The anatomy of fish muscle is different from the anatomy of terrestrial mammals, in that the fish lacks the tendinous system connecting muscle bundles to the skeleton of the animal. Instead, fish has muscle cells running in parallel and connected to sheaths of connective tissue (myocommata), which are anchored to the skeleton and the skin. The bundles of parallel muscle cells are called myotomes.

All muscle cells extend the full length between two myocommata, and run parallel with the longitudinal direction of the fish. The muscle mass on each side of the fish makes up the fillet, of which the upper part is termed the dorsal muscle and the lower part the ventral muscle.

The fillet is heterogenous in that the length of the muscle cells vary from the head end (anterior) to the tail end (posterior). The longest muscle cells in cod are found at about the twelfth myotome counting from the head, with an average length around 10 mm in a fish that is 60 cm long . The diameter of the cells also vary, being widest in the ventral part of the fillet. The myocommata run in an oblique, almost "plow-like" pattern perpendicular to the long axis of the fish, from the skin to the spine. This anatomy is ideally suited for the flexing muscle movements necessary for propelling the fish through the water.

As in mammals, the muscle tissue of fish is composed of striated muscle. The functional unit, *i.e.*, the muscle cell, consists of sarcoplasma containing nuclei, glycogen grains, mitochondria, etc., and a number (up to 1 000) of myofibrils. The cell is surrounded by a sheath of connective tissue called the sarcolemma. The myofibrils contain the contractile proteins, actin and myosin. These proteins or filaments are arranged in a characteristic alternating system making the muscle appear striated upon microscopic examination. Most fish muscle tissue is white but, depending on the species, many fish will have a certain amount of dark tissue of a brown or reddish colour. The dark muscle is located just under the skin along the side of the body.

The proportion of dark to light muscle varies with the activity of the fish. In pelagic fish, *i.e.*, species such as herring and mackerel which swim more or less continuously, up to 48 % of the body weight may consist of dark muscle . In demersal fish, *i.e.*, species which feed on the bottom and only move periodically, the amount of dark muscle is very small.

There are many differences in the chemical composition of the two muscle types, some of the more noteworthy being higher levels of lipids and myoglobin in the dark muscle.

From a technological point of view, the high lipid content of dark muscle is important because of problems with rancidity.

The reddish meat colour found in salmon and sea trout does not originate from myoglobin but is due to the red carotenoid, astaxanthin. The function of this pigment has not been clearly established, but it has been proposed that the carotenoid may play a role as an antioxidant. Further, the accumulation in the muscle may function as a depot for pigment needed at the time of spawning when the male develops a strong red colour in the skin and the female transport carotenoids into the eggs. The latter seems to depend heavily on the amount of carotenoids for proper development after fertilization. It is clearly seen that the muscle colour of salmonids fades at the time of spawning.

The fish cannot synthesize astaxanthin and is thus dependent on ingestion of the pigment through the feed. Some salmonids live in waters where the natural prey does not contain much carotenoid, *e.g.*, in the Baltic Sea, thus resulting in a muscle colour less red than salmonids from other waters. This may be taken as an indication that the proposed physiological function of astaxanthin in salmonids explained above may be less important.

In salmon aquaculture, astaxanthin is included in the feed, as the red colour of the flesh is one of the most important quality criteria for this species.

Muscle contraction starts when a nervous impulse sets off a release of Ca + + from the sarcoplasmic reticulum to the myofibrils. When the Ca + + concentration increases at the active enzyme site on the myosin filament, the enzyme ATP-ase is activated.

This ATP-ase splits the ATP found between the actin and myosin filaments, causing a release of energy. Most of this energy is used as contractile energy making the actin filaments slide in between the myosin filaments in a telescopic fashion, thereby contracting the muscle fibre. When the reaction is reversed (*i.e.*, when the Ca + + is pumped back, the contractile ATP-ase activity stops and the filaments are allowed to slip passively past each other), the muscle is relaxed.

The energy source for ATP generation in the light muscle is glycogen, whereas the dark muscle may also use lipids. A major difference is, further, that the dark muscle contains much more mitochondria than light muscle, thus enabling the dark muscle to operate an extensive aerobic energy metabolism resulting in CO_2 and H_2O as the end products. The light muscle, mostly generating energy by the anaerobic metabolism, accumulates lactic acid which has to be transported to the liver for further metabolization. In addition, the dark muscle is reported to possess functions similar to those are found in the liver.

The different metabolic patterns found in the two muscle types makes the light muscle excellently fitted for strong, short muscle bursts, whereas the dark muscle is designed for continual, although not so strong muscle movements. Post mortem the biochemical and physiological regulatory functions operating in vivo ceases, and the energy resources in the muscle are depleted. When the level of ATP reaches its minimum, myosin and actin are interconnected irreversibly, resulting in rigor mortis.

The Cardiovascular System

The cardiovascular system is of considerable interest to the fish technologist since it is important in some species to bleed the fish (*i.e.*, remove most of the blood) after capture.

The fish heart is constructed for single circulation. In bony fish it consists of two consecutive chambers pumping venous blood towards the gills via the ventral aorta.

After being aerated in the gills, the arterial blood is collected in the dorsal aorta running just beneath the vertebral column and from here it is dispersed into the different tissues via the capillaries. The venous blood returns to the heart, flowing in veins of increasingly larger size (the biggest is the dorsal vein which is also located beneath the vertebral column). The veins all gather into one blood vessel before entering the heart. The total volume of the blood in fish ranges from 1.5 to 3.0 % of the body weight. Most of it is located in the internal organs while the muscular tissues, constituting two- thirds of the body weight, contain only 20 % of the blood volume. This distribution is not changed during exercise since the light muscle in particular is not very vascularized.

During blood circulation the blood pressure drops from around 30 mg Hg in the ventral aorta to 0 when entering the heart . After the blood has passed through the gills, the blood pressure derived from the pumping activity of the heart is already greatly decreased. Muscle contractions are important in pumping the blood back to the heart and counterflow is prevented by a system of paired valves inside the veins. Clearly, the single circulation of fish is fundamentally different from the system in mammals, where the blood passes through the heart twice and is propelled out into the body under high pressure due to the contractions of the heart.

In fish, the heart does not play an important role in the transportation of blood from the capillaries back to the heart. This has been confirmed in an experiment where the impact of different bleeding procedures on the colour of cod fillets was examined. No difference could be found regardless of whether the fish had been bled by means of cutting the throat in front of or behind the heart before gutting, or had not been cut at all before slaughter.

In some fisheries, bleeding of the fish is very important as a uniform white fillet is desirable. In order to obtain this, a number of countries have recommended that fish are bled for a period (15-20 min) prior to being gutted. This means that throat cutting and gutting must be carried out in two separate operations and that special arrangements (bleeding tanks) must be provided on deck. This complicates the working process (two operations instead of one), time-consuming for the fishermen and increases the time-lag before the fish is chilled. Furthermore it requires extra space on an otherwise crowded working deck.

Several researchers have questioned the necessity of handling the fish in a two-step procedure involving a special bleeding period. There seems to be general agreement about the following:

- bleeding is more affected by time onboard prior to bleeding/gutting than by the actual bleeding/gutting procedure.
- best bleeding is obtained if live fish are handled, but it is of major importance to cut the fish before it enters rigor mortis since it is the muscle contractions that force the blood out of the tissues.

Disagreement exists as to the cutting method. Huss and Asenjo found best bleeding if a deep throat cut including the dorsal aorta was applied, but this was not confirmed in the work of Botta *et al.* (1986). The latter also recommended to include a bleeding period (two-step procedure) when live fish were handled (fishing with pound net, trap, seine, longline or jigging), while Valdimarsson *et al.* (1984) found that the quality of dead cod (4 h after being brought onboard) was slightly improved using the two-step procedure. However, it should be pointed out that the effect of bleeding should also be weighted against the advantages of having a fast and effective handling procedure resulting in rapid chilling of the catch.

Discoloration of the fillet may also be a result of rough handling during catch and catch handling while the fish is still alive. Physical mishandling in the net (long trawling time, very large catches) or on the deck (fishermen stepping on the fish or throwing boxes, containers and other items on top of the fish) may cause bruises, rupture of blood vessels and blood oozing into the muscle tissue (haematoma). Heavy pressure on dead fish, when the blood is clotted (*e.g.*, overloading of fish boxes) does not cause discoloration, but the fish may suffer a serious weight loss.

Other Organs

Among the other organs, only the roe and liver play a major role as foodstuffs. Their size depends on the fish species and varies with life cycle, feed intake and season. In cod the weight of the roe varies from a few percent up to 27 % of the body weight and the weight of the liver ranges from 1 to 4.5 %. Likewise, the composition can change and the oil content of the liver vary from 15 to 75 %, with the highest values being found during autumn .

GROWTH AND REPRODUCTION

During growth it is the size of each muscle cell that increases rather than the number of muscle cells. Also, the proportion of connective tissue increases with age. Most fish become sexually mature when they reach a size characteristic of the species and is this not necessarily directly correlated with age. In general, this critical size is reached earlier in males than in females. As the growth rate decreases after the fish has reached maturity, it is therefore often an economic advantage to rear female fish in aquaculture. Every year mature fish use energy to build up the gonads (the roe and milk). This gonadal development causes a depletion of the protein and lipid reserves of the fish since it takes place during a period of low or no food intake.

In North Sea cod it was found that prior to spawning the water content of the muscle increases and the protein content decreases. In extreme cases the water content of very large cod can attain 87 % of the body weight prior to spawning .

The length of the spawning season varies greatly between species. Most species have a marked seasonal periodicity, while some have ripe ovaries for nearly the whole year. The depletion of the reserves of the fish during gonadal development can be extremely severe, especially if reproduction is combined with migration to the breeding grounds. Some species, *e.g.*, Pacific salmon *(Oncorhynchus* spp.), eel *(Anguilla anguilla)* and others, manage to migrate only once, after which they degenerate and die. This is partly because these species do not eat during migration so that, in the case of a salmon, it can lose up to 92 % of its lipid, 72 % of its protein and 63 % of its ash content during migration and reproduction . On the other hand, other fish species are capable of reconstituting themselves completely after spawning for several years. The North Sea cod lives for about eight years before spawning causes its death, and other species can live even longer . In former times, 25-year-old herring *(Clupea harengus)* were not unusual in the Norwegian Sea, and plaice *(Pleuronectes platessa) up* to 35 years old have been found. One of the oldest fish reported was a sturgeon *(Acipenser sturio)* from Lake Winnebago in Wisconsin. According to the number of rings in the otolith, it was over 100 years old.

AQUATIC RESOURCES AND THEIR UTILIZATION

More than two-thirds of the world's surface is covered by water and the total yearly production of organic material in the aquatic environment has been estimated at about 40 000 million t . Tiny microscopic plants, the phytoplankton, are the primary producers of organic material using the energy supplied by the sun.

This enormous primary production is the first link in the food chain and forms the basis for all life in the sea. How much harvestable fish results from this primary production has been the subject of much speculation.

However, there are great difficulties in estimating the ecological efficiency, *i.e.*, the ratio of total production at each successive trophic level. Gulland (1971) reports a range from 10 to 25 % but suggests 25 % as the absolute upper limit of ecological efficiency; for example, not all of the production at one trophic level is consumed by the next.

Ecological efficiency also varies between levels, being higher at the lower levels of the food chain with smaller organisms using proportionally more of their food intake for growth rather than for maintenance. Diseases, mortality, pollution, etc. may also influence ecological efficiency. Since production is greater in the early stages of the food chain, the potential catch is also greater if harvesting is carried out at these stages.

Up to 1970, the world catch of marine fish continued to rise at an overall rate of 6 percent per year, according to FAO statistics. Great optimism was expressed by various authors who estimated the potential world catch to be somewhere between 200 million t/year to 2 thousand t/year ; most of this wide variation being due to uncertainties concerning the trophic level at which the harvest would be taken.

The yearly increase in catches has slowed down since 1970, and the total catch reached a peak of 100 million t in 1989. Since then it has started to drop as a number of fish stocks have begun to collapse, in many cases due to overfishing. However, a slight upward trend is noticed for 1992 and for 1993 world catch is estimated to reach 101 million t. While total catch has started to decline since the peak in 1989, the catch from developing countries as a group is still increasing and since 1985 has exceeded that from developed countries.

Thus in 1992 little more than 60 % of the total world catch was taken by developing countries, and it is estimated that this figure will increase to 66% in 1993. This also means that an increasing part of the world fish catch is taken from warm tropical waters.Are we then reaching the limits of production from "wild" aquatic resources now or do the optimistic predictions from the 1970s still hold? The answer to this question is not only in the affirmative, but for many resources the limit was reached decades earlier than the peak in global landings .

A combination of factors has helped to mark the depletion of many conventional resources. One of these is that continued investments in fishing fleets throughout the world has meant that although catch rates and abundance of high value fish species have often declined, the overall level of fishing effort has increased so that roughly similar levels of landings are being taken at much greater cost to many fishing nations. The real problems with decreasing fish stocks are familiar. First there is "the tragedy of the commons" - whatever lacks a known owner, whether buffalo or fish - which everyone will race to exploit and ultimately destroy.

The next problem which can be identified is the exceptionally poor management of the aquatic resources. What has been done has been too late and too little. The 1982 Law of the Sea, which extended the territorial seas from 12 to 200 miles, gave the coastal States an opportunity to take a protective interest in their fishing grounds.

Instead, many of them rushed to plunder the resources by offering generous subsidies and tax relief for new vessels. Also, the much used quota-system is subject to severe criticism. Often, the net result is increased fishing and increased waste, as perfectly good fish are thrown overboard if quotas are already reached.

Many fish stocks (such as pollack, haddock and halibut off New England) are now considered "commercially extinct"; that is, there are now too few fish to warrant catching. From an initial stage of under-utilization the fishing passes through a phase of rapid expansion until the limit of the resource is reached. This is then followed by a period o overfishing with high fishing effort, but reduced catches until finally - and hopefully - a phase of proper management is reached.

Details on resource management are beyond the scope of this book, but should include the concept of sustainability, environmental aspects and responsible fishing.

However, in an FAO publication it is stated that change from a focus on short-term development of fishing fleets to proper management is a necessary, but insufficient condition for sustainable development. In the same report it is further stated that "Sustainable Development" as promoted at the United Nations Conference on Environment and Development (UNCED) in 1992 cannot be achieved under open-access regimes, whether these are within or outside national territorial waters.

In contrast, the world aquaculture production inclusive of aquatic plants has steadily increased over the last decade totalling 19.3 million ton in 1992, almost half of this (49% is produced in marine aquaculture, 44% in inland aquaculture, and the rest in brackish environment. About 49% of world aquaculture production are fish.

Production of aquati plants is increasing rapidly and reached 5.4 million t in 1992, while smaller increases if production of molluscs and crustaceans are seen. The total value of the aquaculture production is estimated to more than $US 32.5 billion in 1992. To summarize, it can be said that further increases in supply of fish can be expected from better utilization/ reduction of losses and further expansion of aquaculture.

Table shows the breakdown of world fish production.

Table : Breakdown of World Ffish Production
(percentage of world total in live weight)

Year	*For human consumption*			*Other purposes*		
	Total	*Fresh*	*Freezing*	*Curing*	*Canning*	*Animal Feed*
1982	71.1	19.4	25.3	12.8	13.6	28.9
1992	72.8	27.0	24.1	9.3	12.4	27.2

Table shows relatively modest differences in the breakdown of the fish production during the decade 1982-92. However, there was a significant increase in fresh fish consumption. Total fish for human consumption increased by 1.2% while fish used for curing and canning continued to decrease. In value terms, fishery exports reached an estimated $US 40.1 billion in 1993 . Exports of fish and fishery products from developing countries continued to increase reaching a total value of $US 19.4 billion in 1993. In the same year exports from developed countries dropped by 5% to an estimated total value of $US 20.7 billion. Developing countries recorded an increasingly positive trade balance in fish trade, which reached $US 12.7 billion in 1993 .

It should be noted that Table does not give a true picture of the amount of fish available for human food. An enormous amount of fish is wasted due to discards on board or post-harvest losses during processing and distribution. It has been estimated that the global amount of discards is in the range of 17-39 million t/year with an average of 27 million t/year . It has been further estimated that the total post-harvest losses in fish products are about 10 % (James, D., personal communication 1994). These high losses are mainly due to problems of fisheries management, and lack of proper technology and of economic incentives.

THE BACTERIAL FLORA ON LIVE FISH

Microorganisms are found on all the outer surfaces (skin and gills) and in the intestines of live and newly caught fish. The total number of organisms vary enormously and Liston (1980) states a normal range of 10^2–10^7 cfu (colony forming units)/cm^2 on the skin surface. The gills and the intestines both contain between 10^3 and 10^9 cfu/g .

The bacterial flora on newly-caught fish depends on the environment in which it is caught rather than on the fish species . Fish caught in very cold, clean waters carry the lower numbers whereas fish caught in warm waters have slightly higher counts. Very high numbers, *i.e.*, 10^7 cfu/cm^2 are found on fish from polluted warm waters. Many different bacterial species can be found on the fish surfaces. The bacteria on temperate water fish are all classified according to their growth temperature range as either psychrotrophs or psychrophiles. Psychrotrophs (cold-tolerant) are bacteria capable of growth at 0°C but with optimum around 25°C. Psychrophiles (cold-loving) are bacteria with maximum growth temperature around 20°C and optimum temperature at 15°C .

In warmer waters, higher numbers of mesophiles can be isolated. The microflora on temperate water fish is dominated by psychrotrophic Gram-negative rodshaped bacteria belonging to the genera *Pseudomonas, Moraxella, Acinetobacter, Shewanella* and *Flavobacterium*. Members of the *Vibrionaceae (Vibrio* and *Photobacterium)* and the *Aeromonadaceae (Aeromonas* spp.) are also common aquatic bacteria and typical of the fish flora. Gram-positive organisms as *Bacillus, Micrococcus, Clostridium, Lactobacillus* and coryneforms can also be found in varying proportions, but in general, Gram-negative bacteria dominate the microflora. Shewan (1977) concluded that Gram-positive *Bacillus* and Micrococcus dominate on fish from tropical waters.

However, this conclusion has later been challenged by several studies which have found that the microflora on tropical fish species is very similar to the flora on temperate species. A microflora consisting of *Pseudomonas, Acinetobacter, Moraxella* and *Vibrio* has been found on newly-caught fish in several Indian studies . Several authors conclude, as Liston (1980), that the microflora on tropical fish often carry a slightly higher load of Gram-positives and enteric bacteria but otherwise is similar to the flora on temperate-water fish.

Aeromonas spp. are typical of freshwater fish, whereas a number of bacteria require sodium for growth and are thus typical of marine waters. These include *Vibrio, Photobacterium* and *Shewanella*. However, although *Shewanella putrefaciens* is characterized as sodium-requiring, strains of *S. putrefaciens* can also be isolated from freshwater environments. Although *S. putrefaciens* has been isolated from tropical freshwaters, it is not important in the spoilage of freshwater fish. In polluted waters, high numbers of *Enterobacteriaceae* may be found. In clean temperate waters, these organisms disappear rapidly, but it has been shown that *Escherichia coli* and *Salmonella* can survive for very long periods in tropical waters and once introduced may almost become indigenous to the environment .

The taxonomy of S. *putrefaciens* has been rather confused. The organism was originally associated with the *Achromobacter* group but was later placed in the Shewan *Pseudomonas* group IV. Based on percentage of guanine+ cytosine (GC%) it was transferred to the genus *Alteromonas,* but on the basis of 5SRNA homology it was reclassified to a new genus, *Shewanella*. It has recently been suggested that the genus *Aeromonas* spp. which was a member of the *Vibrionaceae* family be transferred to its own family, the *Aeromonadaceae* . Japanese studies have shown very high numbers of microorganisms in the gastrointestinal tract of fish, and as such numbers are much higher than in the surrounding water, this indicates the presence of a favourable ecological niche for the microorganisms. Similarly, Larsen *et al.* (1978) reported up to 10^7 cfu/g of vibrio-like organisms in the intestinal tract of cod and Westerdahl *et al.* (1991) also isolated high numbers of vibrio-like organisms from the intestines of turbot. *Photobacterium phosphoreum* which can be isolated from the surface can also be isolated in high numbers from the intestinal tract of some fish species . On the contrary, some authors believe that the microflora of the gastrointestinal tract is merely a reflection of the environment and the food intake.

Microbial Invasion

The flesh of healthy live or newly-caught fish is sterile as the immune system of the fish prevents the bacteria from growing in the flesh. When the fish dies, the immune system collapses and bacteria are allowed to proliferate freely. On the skin surface, the bacteria to a large extent colonize the scale pockets.

During storage, they invade the flesh by moving between the muscle fibres. Murray and Shewan (1979) found that only a very limited number of bacteria invaded the flesh during iced storage. Ruskol and Bendsen (1992) showed that bacteria can be detected by microscope in the flesh when the number of organisms on the skin surface increases above 10^6 cfu/cm^2. This was seen at both iced and ambient temperatures. No difference was found in the invasive patterns of specific spoilage bacteria and non-spoilage bacteria.

Since only a limited number of organisms actually invade the flesh and microbial growth mainly takes place at the surface, spoilage is probably to a large extent a consequence of bacterial enzymes diffusing into the flesh and nutrients diffusing to the outside.

Fish spoil at very different rates, and differences in surface properties of fish have been proposed to explain this. Skins of fish have very different textures. Thus whiting *(Merlangius merlangus)* and cod *(Gadus morhua)* which have a very fragile integument spoil rapidly compared to several flatfish such as plaice that has a very robust dermis and epidermis. Furthermore, the latter group has a very thick slime layer, which includes several antibacterial components, such as antibodies, complement and bacteriolytic enzymes.

SPECIFIC SPOILAGE ORGANISMS

Bacteria on fish caught in temperate waters will enter the exponential growth phase almost immediately after the fish have died. This is also true when the fish are iced, probably because the microflora is already adapted to the chill temperatures. During ice storage, the bacteria will grow with a doubling time of approximately 1 day and will, after 2-3 weeks, reach numbers of 10^8-10^9 cfu/g flesh or cm^2 skin. During ambient storage, a slightly lower level of 10^7-10^8 cfu/g is reached in 24 hours. The bacteria on fish caught in tropical waters will often pass through a lag-phase of 1-2 weeks if the fish are stored in ice, whereafter exponential growth begins. At spoilage, the bacterial level on tropical fish is similar to the levels found on temperate fish species. If iced fish are stored under anaerobic conditions or if stored in CO_2 containing atmosphere, the number of the normal psychrotrophic bacteria such as S. *putrefaciens* and *Pseudomonas* is often much lower, *i.e.*, 10^6-10^7 cfu/g than on the aerobically stored fish. However, the level of bacteria of psychrophilic character such as P. *phosphoreum* reaches a level of 10^7-10^8 cfu/g when the fish spoil.

The composition of the microflora also changes quite dramatically during storage. Thus, under aerobic iced storage, the flora is composed almost exclusively of *Pseudomonas* spp. and S. *putrefaciens* after 1-2 weeks. This is believed to be due to their relatively short generation time at chill temperatures and is true for all studies carried out whether on tropical or temperate-water fish.

At ambient temperature (25°C), the microflora at the point of spoilage is dominated by mesophilic *Vibrionaceae* and, particularly if the fish are caught in polluted waters, *Enterobacteriaceae.* A clear distinction should be made between the terms spoilage flora and spoilage bacteria since the first describes merely the bacteria present on the fish when it spoils whereas the latter is the specific group that produce the off-odours and off-flavours associated with spoilage. A large part of the bacteria present on the spoiled fish have played no role whatever in the spoilage. Each fish product will have its own specific spoilage bacteria and the number of these will, as opposed to the total number, be related to the shelf life.

It is not an easy task to determine which of the bacteria isolated from the spoiled fish are those causing spoilage, and it requires extensive sensory, microbiological and chemical studies. First, the sensory, microbiological and chemical changes during storage must be studied and quantified, including a determination of the level of a given chemical compound that correlates with spoilage (the chemical spoilage indicator). Second, bacteria are isolated at the point of sensory rejection. Pure and mixed cultures of bacteria are screened in sterile fish substrates for their spoilage potential, *i.e.*, their ability to produce sensory (off-odours) and chemical changes typical of the spoiling product. Finally, the selected strains are tested to evaluate their spoilage activity, *i.e.*, if their growth rate and their qualitative and quantitative production of off-odours are similar to the measurements in the spoiled product. The latter step is particularly important, as some bacteria may produce the chemical compounds associated with spoilage but are unable to do so in significant amounts, and they are thus not the specific spoilage bacteria. When stored aerobically, levels of 10^8-10^9 cfu/g of specific spoilage bacteria are required to cause spoilage. The spoilage of packed fish is seen at a much lower level of 10^7 cfu P. *phosphoreum* per gramme. This relatively low level is probably due to the very large size (5 μm) of the bacterium resulting in a much higher yield of for example, TMA per cell .

SHEWANELLA PUTREFACIENS

Spoilage potential and activity can be assessed in several fish substrates as sterile, raw fish juice, heat-sterilized fish juice or on sterile muscle blocks . The latter is the most complicated but is also that yielding results comparable to the product. If any of the fish juices are chosen, it is important that the growth rate of the spoilage bacteria in the model system is equal to the growth rate in the product. A qualitative test for the ability of the bacteria to produce H_2S and/or reduce TMAO may also be used when the spoilage flora is screened for potential spoilage bacteria. A medium where the reduction of TMAO to TMA is seen as a redox indicator changes colour, and the formation of H_2S is evident from a black precipitation of FeS which has been developed for this purpose .

Shewanella putrefaciens has been identified as the specific spoilage bacteria of marine temperate- water fish stored aerobically in ice. If the product is vacuum-packed, P. phosphoreum participates in the spoilage and it becomes the specific spoilage bacteria of CO_2 packed fish. The spoilage flora on iced tropical fish from marine waters is composed almost exclusively of *Pseudomonas* spp. and *S. putrefaciens*. Some *Pseudomonas* spp. are the specific spoilers of iced stored tropical freshwater fish and are also, together with *S. putrefaciens,* spoilers of marine tropical fish stored in ice. At ambient temperature, motile aeromonads are the specific spoilers of aerobically stored freshwater fish. Barile *et al.* (1985) showed that a large proportion of the flora on ambient-stored mackerel consisted of *S. putrefaciens,* indicating that this bacterium may also take part in the spoilage.

(1) Modified Atmosphere Packaging (CO_2 containing)
(2) LAB: Lactic Acid Bacteria
(3) Fish caught in tropical waters or freshwaters tend to have a spoilage dominated by *Pseudomonas spp.*

BACTERIAL GROWTH DURING STORAGE AND SPOILAGE

These include trimethylamine, volatile sulphur compounds, aldehydes, ketones, esters, hypoxanthine as well as other low molecularweight compounds. The substrates for the production of volatiles are the carbohydrates (*e.g.*, lactate and ribose), nucleotides (*e.g.*, inosine mono-phosphate and inosine) and other NPN molecules. The amino-acids are particularly important substrates for formation of sulphides and ammonia. Microorganisms obtain far more energy from aerobic oxidation than from an anaerobic fermentation; thus the complete oxidation of 1 mole glucose (or other hexose) via Kreb's cycles yields 6 moles of CO_2 and 36 moles of ATP. On the contrary, the fermentation of 1 mole glucose gives only 2 moles of ATP and two moles of lactic acid. The initial aerobic growth on fish is dominated by bacteria using carbohydrates as substrate and oxygen as terminal electron-acceptor with the concurrent production of CO_2 and H_2O.

Reduction of Trimethylarnine Oxide (TMAO)

The growth of oxygen-consuming bacteria results in the formation of anaerobic or microaerophilic niches on the fish. This does, however, not necessarily favour the growth of anaerobic bacteria. Some of the bacteria present on fish are able to carry out a respiration (with the ATP advantage) by using other molecules as electron acceptor. It is typical of many of the specific spoilage bacteria on fish that they can use TMAO as electron acceptorin an anaerobic respiration. The reduced component, TMA, which is one of the dominant components of spoiling fish, has a typical fishy odour. The level of TMA found in fresh fish rejected by sensory panels varies between fish species, but is typically around 10-15 mg TMA-N/100 g in aerobically stored fish and at a level of 30 mg TMA-N/100 g in packed cod. The TMAO reduction is mainly associated with the genera of bacteria typical of the marine environment *(Alteromonas, Photobactetium, Vibrio* and S. *putrefaciens),* but is also carried out by *Aeromonas* and intestinal bacteria of the *Enterobacteriaceae.* TMAO reduction has been studied in fermentative, facultative anaerobic bacteria like *E. coli* and *Proteus* spp. as well as in the non-fermentative S. *putrefaciens*.

During aerobic growth, S. *putrefaciens* uses the Kreb's cycle to produce the electrons that are later channelled through the respiratory chain. Ringo *et al.* (1984) suggested that during anaerobic respiration S. *putrefaciens* also uses the complete Kreb's cycle, whereas it has recently been shown that in the anaerobic respiration in S. *putrefaciens,* only part of the Kreb's cycle is used and electrons are also generated by another metabolic pathway, namely the serine pathway . S. *putrefaciens* can use a variety of carbon sources as substrate in its TMAO-dependent anaerobic respiration, including formate and lactate. Compounds like acetate and succinate that are used in the oxygen respiration cannot be used when TMAO is terminal electron acceptor and on the contrary, acetate is a product of the anaerobic TMAO reduction.

Contrary to this, sugars and lactate are the main substrates generating electrons when *Proteus* spp. reduces TMAO. The reduction is accompanied by a production of acetate as the main product . TMAO is, a typical component of marine fish, and it has recently been reported that also some tropical freshwater fish contain high amounts of TMAO . However, TMA is not necessarily a characteristic component during spoilage of such fish because spoilage is due to *Pseudomonas spp.* .

Production of hypoxanthine

The development of TMA is in many fish species paralleled by a production of hypoxanthine. Hypoxanthine can be formed by the autolytic decomposition of nucleotides, but it can also be formed by bacteria; and the rate of bacterial formation is higher than the autolytic. Both Jorgensen *et al.* (1988) and Dalgaard (1993) showed a linear correlation between the contents of TMA and hypoxanthine during iced storage of packed cod. Several of the spoilage bacteria produce hypoxanthine from inosine or inosine monophosphate, including *Pseudomonas* spp. S. *putrefaciens* and P. *phosphoreum* . In cod and other gadoid fishes, TMA constitutes most of the so-called total volatile bases, TVB (also called total volatile nitrogen, TVN) until spoilage. However, in the spoiled fish where the TMAO supplies are depleted and TMA has reached its maximum level, TVB levels still rise due to formation of NH3 and other volatile amines. A little ammonia is also formed in the first weeks of iced storage due to autolysis. In some fish that do not contain TMAO or where spoilage is due to a non-TMAO reducing flora, a slow rise in TVB is seen during storage, probably resulting from the deamination of amino-acids.

Volatile sulphur-compounds are typical components of spoiling fish and most bacteria identified as specific spoilage bacteria produce one or several volatile sulphides. S. *putrefaciens* and some *Vibrionaceae* produce H_2S from the sulphur containing amino-acid 1-cysteine. On the contrary, neither *Pseudomonas* nor P. *phosphoreum* produce significant amounts of H_2S. Thus, hydrogen sulphide, which is typical of spoiling iced cod stored aerobically, is not produced in spoiling CO_2 packed fish . Methylmercaptan (CH_3SH) and dimethylsulphide ($(CH_3)_2S$) are both formed from the other sulphur-containing amino-acid, methionine. Taurine, which is also sulphur-containing, occurs as free amino-acid in very high concentrations in fish muscle. It disappears from the fish flesh during storage but this is because of leakage rather than because of bacterial attack .

The volatile sulphur-compounds are very foul-smelling and can be detected even at ppb levels, so even minimal quantities have a considerable effect on quality. Ringo *et al. (1984)* have shown that cysteine is used as substrate in the Kreb's cycle when electrons are transferred to TMAO, and the formation of H_2S and TMA is thus to some extent a linked reaction. Contrary to the iced spoilage by S. *putrefaciens* and the ambient spoilage by *Vibrionaceae* which is dominated by H_2S and TMA, the spoilage caused by *Pseudomonas* spp. is characterized by absence of these compounds.

Table : Typical spoilage compounds during spoilage of fresh fish stored aerobically or packed in ice or at ambient temperature

Specific spoilage organism	Typical spoilage compounds
Shewanella putrefaciens	TMA, H_2S, CH_3SH, $(CH_3)_2S$, Hx
Photobacterium phosphoreum	TMA, Hx
Pseudomonas spp.	ketones, aldehydes, esters, non-H_2S sulphides
Vibrionaceae	TMA, H_2S
anaerobic spoilers	NH_3, acetic, butyric and propionic acid

PSEUDOMONAS SPOILAGE OF ICED FISH

Fruity, rotten, sulphydryl odours and flavours are typical of the Pseudomonas spoilage of iced fish. *Pseudomonas* spp. produce a number of volatile aldehydes, ketones, esters and sulphides. However, it is not known which specific compounds are responsible for the typical off odours. The fruity off-odours produced by *Pseudomonas fragi* originate from monoaminomonocarboxylic amino acids. TVB will continue to rise even after TMA has reached its maximum. This latter rise is due to proteolysis commencing when several of the free amino-acids have been used. Lerke *et al.* (1967) separated fish juice into a protein and a non-protein fraction and inoculated spoilage bacteria in each fraction and in the whole juice. The non-protein fraction of a fish juice spoiled as the whole juice whereas only faint off-odours were detected in the protein fraction of the juice. Although some authors have used the number of proteolytic bacteria as indicators of spoilage, it must be concluded that the turnover of the protein fraction is not of major importance in spoilage of fresh fish.

Some of the compounds typically formed by bacteria during spoilage of fish are shown in Table below together with the substrate used for the formation. The formation of TMA is accompanied by a formation of ammonia during anaerobic storage of herring and mackerel . Prolonged anaerobic storage of fish results in vigorous production of NH_3 owing to further degradation of the amino-acids, and in the accumulation of lower fatty acids as acetic, butyric and propionic acid. The very strong NH_3-producers were found to be obligate anaerobes belonging to the family Bacteroidaceae genus *Fusobacterium*. These organisms grow only in the spoiled fish extract and have little or no proteolytic activity relying on already hydrolysed proteins.

Table : Substrate and off-odour/off-flavour compounds produced by bacteria during spoilage of fish

Substrate	Compounds produced by bacterial action
TMAO	TMA
cysteine	H_2S
methionine	CH_3SH, $(CH_3)_2S$
carbohydrates and lactate	acetate, CO_2, H_2O
inosine, IMP	hypoxanthine
amino-acid s	
(glycine, serine, leucine)	esters, ketones, aldehydes
amino-acids, urea	NH_3

During iced storage of fresh fatty fish, changes in the lipid fraction is caused almost exclusively by chemical action, *e.g.*, oxidation, whereas bacterial attack on the lipid fraction contributes little to the spoilage profile. During storage of lightly preserved fish, lipid hydrolysis caused by bacteria may be part of the spoilage profile.

3

Integrated Farming in Fisheries Sector

INTEGRATED FISH FARMING

Integrated fish farming, an efficient method of combining fisheries, agriculture and livestock to increase the overall production and obtain maximum utilization of resources is a traditional practice in China. High fish production has been achieved. In order to transfer and adapt the Chinese technology in developing countries having different agro-climatic conditions, there is a need to establish through research a better understanding of the biological processes that are taking place in fish ponds under the system of integrated fish farming. The fish pond is often treated as a "black box". We stock fish; we add manures and grasses and in some cases chemical fertilizers or processed feeds. Via photosynthesis algae grow. In several months, if we have managed the pond well, marketable fish are harvested. What the actual sources of the fish growth were we can only estimate from analyses of the gut content. But such analyses reveal only the foods most recently consumed and almost always the major component is "unidentified organic matter".

A tracer is needed that will show incorporation of food into the fish's body. Radioactive tracers (^{14}C: ^{32}P) are useable only in very small ponds or tanks. Even then the feeds must be artificially labelled with the radioactive isotope.

There is in nature a naturally occurring tracer that is present in all organic matter. It is the ratio of the 2 stable isotopes of carbon (^{13}C: ^{12}C). This ratio, reported as delta C, varies among plant species in a way dependent upon the photosynthetic pathway of the particular plant. The delta C of animal bodies is similar to the delta C of the foods they assimilated. Hence if a sufficiently large range of delta C values exists among the available foods, it is often possible to estimate which foods or groups of foods were assimilated by an animal merely by comparing the delta C of that animal with the delta C of the available foods. Clearly there are limitations to the method. We are using one tracer in a multidimensional system. When several foods are available, it may be possible only to distinguish between 2 groups of foods, each group containing several possible foods with similar delta C values. Even with this limitation the delta C isotope method has given insights into complex food webs.

BIOLOGY OF MAJOR CULTIVATED FISHES

China has a vast area of water, with a distribution of inland rivers, lakes, reserviors and ponds throughout its territory, and hence it is rich in fishery resources. There is an ample variety of fishes, of which more than 50% are carps. Up until now, there have been about 20 species cultured in ponds in integrated fish farms. The introduction here focuses on the four main well-known carps: Silver carp, Bighead, Grass carp and Black carp, and other good stocks———Common carp, Crucian carp, Chinese bream (Wuchang fish), Mud carp and Tilapia etc.

Understanding the habits of fish, mastering the laws of their growth, development, propagation and feeding and satisfying their ecological requirements accordingly will be of great practical significance to the development of fishery production, to the application of farming techniques and to the increase of fish yield.

Morphology and Silver Carp (Hypophthalmichthys molitrix)

Silver carp belong in taxonomy to Family Cyprinidae, Subfamily Hypophthalmichthyinae. Body: compressed Scales: small, Mouth: in front, with lower jaw slightly slanting upward. Eyes: comparatively small, situated below the horizontal axis of body. Gill membrane: unconnected to isthmus. Gill rakers: dense, interlaced and connected, covered with a sponge-like membraneous sieve. Abdominal keel: extending from the base of pectoral fins to the anus. Pectoral fin: its terminal tip does not exceed the base of ventral fin. Pharyngeal teeth: one row in 4/4, with fine lines and tiny grooves on surface. Intestinal length: 6—10 times that of body length. Colour of body: silvery white while alive; colour of dorsal, very dark brown. The largest body so far discovered is about 20 kg.

Big Head (Aristichthys nobilis)

Bighead are similar to Silver carp in shape. They belong to the same genus Hypophthalmichthys. Head: bigger. Snout: short and blunt. Eyes: small, situated below the horizontal axis of body. Gill membrane: unconnected to isthmus. Gill rakers, dense and separated; without spongelike sieve. Abdominal keel: between the bases of ventral fins and the anus. Pectoral fin: the terminal tip reaches 1/3—2/5 of the base of ventral fin. Pharyngeal teeth: one row in 4/4, surface flat. Intestinal length: about 5 times that of body length.

Colour of body: while alive, dorsal and upper sides, light black, scattered with irregular yellowish black spots; ventral surface, silvery white. The largest body so far found is about 40 kg.

Grass Carp (Ctenopharyngoden idellus)

Grass carp are larger-size fish in sub-family Leuciscinae of family Cyprinidae. Body shape: almost cylindric, with flat head and round abdomen. Scales: big. Mouth: in front; lower jaw, shorter. Gill membrane: connected to isthmus. Gill raker: small and short, in scattered arrangement. Pharyngeal teeth: 2 rows in 2,5/4,2 compressed like combs. Intestinal length: 2.3—3.3 times that of body length. Colour of body: while alive, dorsal, grey; abdomen, light grey; sides, greenish yellow; fins, a lighter colour. The largest body so far found is about 35 kg.

Black Carp (Mylopharyngodon piceus)

Black carp are more close to Grass carp among Cyprinidae. Body shape: like Grass carp with a pointed head. Scales: big and circular. Mouth: arc-shaed in front.

Eyes: medium size, situated in the middle part of head sides. Gill raker: short. Gill membrane: connected to the isthmus. Pharyngeal teeth: one row in 5/4, big, short and molar-like; surface smooth. Intestinal length: 1.2—2 times that of body length. Colour of body: while alive, greenish black; dorsal darker; abdomen light grey; Fins, black. The largest body so far discovered is about 70 kg.

Crucian Carp (Carassius auratus)

Crucian carp are very close to Common carp. Body: compressed and relatively thick. Abdomen: round. Head: small and short. Snout: blunt. Mouth: arc-shaped in front. Lip: thick without barbels. Pharyngeal teeth: compressed, one row in 4/4. Intestinal length: 2.7—3.2 times that of body length, some even reaching 5 times.

Colour of body: silvery grey when alive; darker on dorsal and lighter on abdomen. The largest body found is about 1.5 kg. They have a wide distribution and a strong adaptability. They can live in different water body such as rivers, lakes, ponds and ditches with some variations and differentiations in characteristics.

Carassius auratus gibelio are sub-species of Crucian carp. Body: higher than usual Crucian carp. Abdomen: round. Colour of body: bluish grey with silvery white on abdomen.

The largest body is about 3 kg. Japanese Crucian carp originated from Lake Pipa in Japan and was introduced into China in 1976. They grow faster than Carassius auratus gibelio. A large variety of golden fish are bred out through long period of artificial breeding. Chinese bream, or Wuchang fish

(Megalobrama amblycephala) Chinese bream belong to subfamily Abramidinae. Body: high, compressed and lozenge-shaped.

Head: small and short Mouth: slanting. Abdominal keel: extending from the base of pelvic fin to the anus. Pharyngeal teeth: 3 rows arranged in 2,4,5/4,4,2. Intestinal length: 2.7 times that of body length. Colour of body: dark grey while alive, darker on dorsal; scale, dark grey in the middle and lighter on its edge. The largest body so far caught, is about 3 kg.

Common Carp (Cyprinus carpio)

Body: compressed. Dorsal: projected in arch-shape. Abdomen: round. Mouth: slightly downwards with a long blunt snout and with two pairs of barbels on upper jaw; lower pair a little longer. Dorsal fin: long.

Scales: big and thick. Pharyngeal teeth: 3 rows in 1,1,3/3,1,1; teeth on inner sides, molar like. Intestinal length: 1.5—2 times that of body length.

Colour of body: while alive, varying with different living conditions, usually dark grey or yellowish brown on dorsal; sides golden yellow; lower part of caudal fin, red. The largest body so far discovered is about 40 kg.

China has a long history of culturing Common carp, which have a wide distribution and strong adaptability. There are a lot of morphological variations through artificial breeding and natural selections, such as Scale carp, Mirror carp, Wu Yuan red purse carp, Xing Guo red carp etc.

Mud carp (Cirrhina molitorella)

Mud carp belong to family Cyprinidae, subfamily Barbinae, genus Cirrhina. Body: long and compressed. Abdomen: round and slightly flat. Snout: short, round and blunt. Mouth: inferior and transverse, with two pairs of barbels; snout barbels, strong and thick; jaw barbels, small and short.

Caudal fin: deeply separated with upper part a little longer than the lower one. Lateral line: from the upper part of the pectoral fin around the lateral line, there are 8—12 scales with dark dots at their bases which form lozenge-shaped spots.

Fins: dark-grey. Pharyngeal teeth: 3 rows in 2,4,5/5,4,2. Intestinal length: about 14 times that of body length. The largest body caught is about 4 kg.

Tilapia

Tilapia belong to order Perciformes, family Cichlidae, genus Tilapia. This genus comprises more than 100 species including subspecies. At present, 15 of them are taken as cultured fishes all over-the world. In China, mainly Tilapia mossambica and T. nilotica etc. are reared.

Body of T. mossambica: short and compressed. Dorsal: a little higher. Shape: similar to Crucian carp. Mouth: bigger. Lip: thick, with lower jaw a little long than upper jaw. Scales: circular. Lateral line: disjointed. The intestinal length about 7 times that of body length. Colour of body: dark grey while alive; during spawning stage, body colour of male fish, dark green; edges of dorsal, anal and caudal fins, obvious red; female, greyish yellow. The largest body weight so far discovered is about 0.5 kg.

Colour of T. nilotica: changing with external conditions, light black; abdomen white, with 9 longitudinal black stripes on body surface, of which, 7 below dors; fins and 2 on the peduncle; caudal fin with 10 clear vertical black stripes for life. Scales: ctenoid. The largest body weight is 2.5 kg.

Nuptial colour can be found on the male and the female of Tilapia at reproductive period. But their external genital organs are different in appearance.

The male has two pores; anus in front and cloaca in the rear. The female, 3 pores *i.e.* anus in front, genital pore in the middle and excretory pore in the rear.

FEEDING HABITS OF FISH

Although Silver carp, Bighead, Grass carp, Black carp, Common carp and Mud carp, belong to Cyprinidae, they have formed their respective ways of food in-take and their respective food chains at different stages of the development, It is due to their long-term adaptations to ecological conditions, the restrictions of water body environment and the perfection of body structure as well as the development of foodintake and digestive organs along with their growth.

The main natural food Silver carp and Bighead feed on are plankton. Silver carp mainly feed on zooplankton at larval stage, whereas their feeding turns to phytoplankton after grown-up. Bighead feed on zooplankton mainly all their life. The differentiation of feeding habits are caused by the structure and the density of their filtering organs gill rakers. The gill rakers of Silver carp and Bighead are situated in the operculum with 4 pairs on either side, and the 5th pair of gill arch which was specialized into inferior pharyngobranchial. The gill arch bone is the base to which gill rakers and gill filaments attach.

The gill rakers of Bighead are delicate and sabre-shaped, each consisting of a neck, a stem and a base. The neck is narrow and short. The stem is the principal part of the gill raker. The dorsal part of it is thicker, while the ventral part thinner. On either side, there is a row of wartlike protuberance, so to speak, lateral protuberance.

The lateral protuberance of a gill-raker is interwoven with the next one in serration. Occasionally there are opposite protuberances, their triangle-shaped bases cling to the gill arch bones. Besides, gill rakers can be divided into two kinds, broad and narrow ones. There exists one broad gill raker every 3—6 narrow gill rakers. The gill rakers of Silver carp are different from that of Bighead. The gill rakers are inter-connected by minute bony bridges, which are covered with spongy sieve membranes. The density of gill rakers is higher than that of Bighead's gill raker.

The newly hatched Silver carp and Bighead fry nourish themselves with egg yolk and begin to take in plankton in the water when their body length reaches 7—9 mm, 3—4 days after hatching. The food-filtering organs are imperfect with short and sparse gill rakers before the body length reaches 15 mm. At that time, both Silver carp and Bighead eat the same food.

The major groups of zooplankton that Silver carp and Bighead eat are rotifera, nauplius of copepods and tiny cladocera. The shape and structure of filtering organs of fry about 20–30 mm in body length, are generally the same as those of adult fish's. The fry of Silver carp will take in phytoplankton as its main foods, since it has nearly 200 gill rakers, 1 mm in length each, with minute bony bridges between them Sieve membranes covering the bridges form a fine. The gill rakers of Bighead fry, short and sparse, are separated with larger spaces and difficult to detain tiny phytoplankton. As a result, their main food changes from tiny zooplankton to all sorts of zooplankton.

Beside the filtering organs———gill rakers, Silver carp and Bighead have accessary organs———palatine folds, that is, inter-gill raker folds. Palatine folds, located at the apex of mouth cavity, consist of 9 vertical ridges of mucus membrane, four on either side and one in between. The middle ridge is very short like an inverted letter Y with the branched end backwards. Palatine folds act in coordination with gill rakers, to get the food through filtering.

Under culture conditions, Silver carp and Bighead can also feed on commercial feeds such as cakes, brans, dregs and wastes.

Feeding Habits of Grass Carp and Black Carp

The fry of both fishes under 15 mm in body length have the same feeding habits, mainly feeding on zooplankton. But the feeding habits of the fry above 20–30 mm in body length will differentiate. The fry of Grass carp start to take in tender aquatic plants, while the fry of Black carp eat Benthos such as snails and corbicula spp. etc. Grass carp are typical herbivorous species taking in all sorts of aquatic and land grasses, hence, the pharyngeal teeth are well developed, tough and strong (tooth formula: 2,5/4.2). The teeth are shaped like choppers with saw-toothed edges. Pharyngeal teeth at both sides are interlaced with each other.

They are against callous pad of the basioccipital, grinding food into pieces for digestion in intestines. Grass carp are voracious eaters, feeding amount usually could reach 40%. But the plant cellulose could not be digested and utilized. Black carp are carnivorous fish, and usually feed on molluscs like snails, clams and corbicula spp. By using pharyngeal teeth and callous pad, they crush hard shells, then swallow the meat. Their pharyngeal teeth are strong and tough and molar-shaped (tooth formula: 4/5).

Under culture conditions, both fishes tend to be more or less omnivorous, feeding on oil seed cakes, brans, dregs and animal feeds like silkworm pupae, earthworm, animal entrails.

FEEDING HABITS OF COMMON CARP AND CRUCIAN CARP

At the larval stage, feeding habits of the two are basically similar. They chiefly feed on rotifer, cladoceran, copepoda, chironomid larvae and other insects' larvae. Common carp and Crucian carp about 50 mm in body length tend to be omnivorous. Common carp are inclined to be more carnivorous, whereas Crucian carp more herbivorous. Both have feeding habits of phagotrophy. The pharyngeal teeth of Common carp are relatively developed, molarshaped in 3 rows, with transverse grooves on the rest of the inner row except the first tooth, which is smooth. Common carp display a wide adaptation to foods. Their common natural foods are benthos such as snails, young clams, corbicula spp. cladocera, copepod, chironomid larvae, shrimps and insects' larvae. Also they take certain amount of detritus of higher aquatic plants and plant seeds.

The nasal bone of Common carp is well developed so that their premaxilla and mandible could be projected out like a tube to dig the mud for organic detritus. Foods for Crucian carp are chiefly large amounts of detritus, diatom, filament algae, aquatic grasses and plant seeds, including a certain amount of cladocera, copepod, chironomid larvae and water earthworm etc. Under culture conditions, Common carp and Crucian carp also like to take commercial feeds such as oil seed cakes, brans, crops and silkworm pupae, etc.

FEEDING HABITS OF CHINESE BREAM WUCHANG FISH

At larval stage, Chinese bream feed mainly on zooplankton such as cladocera and copepod, whereas, at adult stage, mainly on aquatic grass such as Vallisneria spiralis, Hydrilla verticillata, and Potamogeton malainus and secondarily on Potamogeton crispus, Myriophyllum spicatum, Spirogyra and plant detritus. With a mouth small in size, and the pharyngeal teeth and callous pad small and weak, the ability and intensity of food intake of Wuchang fish fail to come up with Grass carp.

Feeding Habits of Mud Carp

With a small mouth, transverse in inferior position, under the natural conditions, mud carp use the bony edges of upper and lower jaws to scrape diatom, green algae chlamydomonas and filament algae etc. attached to stones and something else. Not seldom, they take detritus of higher plants, bottom humus and a little zooplankton as feeds. Under culture conditions, they like to eat commercial feeds such as seedcakes, dregs, brans, and animal manures.

Feeding Habits of Tilapia

Tilapia are omnivorous fish, with a tendancy to be herbivorous. Tilapia feed mainly on zooplankton at larval stage. The scope of food enlarges with the growth. The usual common foods are all kinds of planktonic, benthic and epiphytic algae, tender higher aquatic plants, all organic detritus and a little animal feeds such as earthworms, small shrimps and aquatic insects. Tilapia nilotica, possessing denser gill rakers (24—31), are more likely to feed on phytoplankton and could utilize the green algae Chlorophyta and blue algae Cyanophyceae which can not be digested by other fishes, but detritus occupies greater proportion among the natural feeds for Tilapia mossambica because they only have 14— 19 gill rakers.

Under culture conditions they can utilize all kinds of vegetable leaves tender grass, animal manures brans, seed cake and pelleted feeds if applied. At a proper temperature, they ability and intensity of their food intake are much higher.

GROWTH RATE AND EVALUATE PRODUCTION EFFICIENCY

The growth rate is an important criterion to evaluate production efficiency.

Silver carp, Bighead, Grass carp, Black carp and Common carp, with the features of large size and speedy growth, have been taken as the dominant cultured species for polyculture in Chinese integrated fish farms, whereas Crucian carp, Wuchang fish and Mud carp, owing, to smaller size and slower growth are usually regarded as a secondary species for polyculture so as to raise per-unit fish yield.

As a mass production, the latter are well-known species for high yields.

The growth rate of every kind of fish involves genetics of species, besides, it's also closely related to water quality, water temperature, nourishment, stocking density and management. As a rule, the growth of body in length and weight is absolutely the fastest before the first sexual maturity, then it will slow down and even stop.

GROWTH OF SILVER CARP AND BIGHEAD

It is the second year that Silver carp grow the fastest. After the fourth year the growth rate decreases during th period of the 3rd — — 6th years, Silver carp gain the maximum weight.Bighead have the fastest growth during the period of the 2nd — — 3rd year, and gain the maximum growth in weight in the 3rd year.

Growth of Grass carp and Black carp

Grass carp have the fastest increase in length during the period of the 1st — — 2nd years, and gain the maximum growth in weight during the period of the 2nd — — 3rd years. Black carp have the fastest increase in length during the period of the 1st — — 2nd years, and gain the maximum growth in weight during the period of the 3rd — — 4th years.

Growth of Common carp, Crucian carp and Wuchang fish Common carp have the fastest increase in length during the period of the 1st — 2nd years.

The female fish grow faster than the male fish, and Common carp gain the maximum growth in weight during the period of the 4th – 5th years.

Crucian carp, have the fastest increase in length in the first year and gain the max growth in weight during the period of the 4th – 5th years.

Wuchang fish have the fastest growth during the period of 1st – 2nd years, afterwards, the growth rate decreases gradually, and they gain the max growth in weight in the second year.

Table : Relationship between the age and the growth of main cultured fish

	Species	Silver carp	Bighead	Grass carp	Black carp	Common carp	Crucian carp	Wuchang fish
age								
item								
max increase in length		2	2--3	1--2	1--2	1--2	1	1--1
max growth in weight		3--6	3	2--3	3--4	4--5	4--5	2

Growth of Mud Carp

Mud carp have the fastest increase in length but gain the minimum growth in weight in the 1st year. The increase in length will slow down gradually. Sexual maturity begins in the 3rd year with the continuation of gaining weight.

Table : Growth rate of Mud carp reared in ponds

	age	1	2	3	4	5	6
item							
body	absolute measurement	14.5	23.4	26.1	29.2	33.6	36.5
length (cm)	annual increase	14.5	8.9	2.7	3.1	4.4	2.9
body	absolute measurement	70.9	257.9	354.5	499.1	739.3	969.5
weight (g)	annual increase	70.9	187.0	96.6	144.6	240.2	230.2

NATURAL REPRODUCTION OF FISH

The natural spawning grounds of Silver carp, Grass carp and Black carp are vastly distributed in Pearl River, Qiantangjiang River, Changjiang River, Huaihe River, northwards up to Heilongjiang River systems, while those of Bighead mainly in Changjiang River, Huaihe River and Pearl River. Mud carp, as a subtropic species have the spawning-ground distribution in the southern part of China like Hainan Island, Guangdong, Guangxi, Fujian and Yunnan. Among all these river systems, the spawning grounds of Changjiang River and Pearl River are the biggest. Quantities of fry of Silver carp, Bighead, Grass carp and Black carp proliferate in Changjiang River & Pearl River systems. Chinese carps usually like to stay in the lower reaches of rivers, or river branches or lakes for fattening, where the water current is slow and the water is fertile with abundance of food. When the spawning season is drawing near, the spawning schools begin to gather schooling and to·migrate towards the middle and upper course spawning grounds. At that time, the gonads of the broodfish have, in a great majority of cases, reached Stage IV while the gonads of the male fish have reached Stage V. If the ecological conditions in the spawning grounds are suitable for reproduction, they will proceed to spawn.

As for the main cultivated fishes in China, the spawning time in different river systems varies with the climatic conditions, but all in the summer. In Changjiang River drainage, Silver carp and Grass carp generally start spawning in late April to early May and Bighead in mid or late May, whereas in Pearl River Drainage, the spawning season begins in mid or late April, a little later for Bighead. As a rule, the North is later than the South by 1—2 months. After July, the fry production is apparently reducing. During this period, the matured brood fish are ready to spawn in the spawning places when the water level of rivers rises and the temperature goes above 18°C below 30°C. Anyhow, the optimum temperature for spawning is 22°C—28°C. As a subtropic species, Mud carp need a little higher temperature for spawning with the optimum temperature of 26°C—30°C, starting from late May through August and September.

Stimulation of Proper Ecological Conditions

Under the stimulation of proper ecological conditions (generally for about 20 hours). Spawning behaviours could be seen when the spawning schools of Silver carp and Grass carp arrive in spawning grounds. There are two common forms ——— on water surface and under water. Spawning on water surface is termed "floating spawning" and the latter "muffled spawning". In "floating spawning", the estrus could be seen that males chase females excitedly, often bumping against the abdomen of the female with their head in swift swimming, jumping out of the water, then splashing in waves. Sometimes, the male and female would float on their backs with their pectoral fins vibrating violently.

When it reaches the climax of the estrus, the female & male discharge eggs and milt respectively. The eggs will be fertilized in the water. As for "muffled spawning", the action would take place under the water. Occasionally waves aroused by chasing could be seen on the surface. More often than not, the male fish are the overwelming majority in spawning schools . For Silver carp the male are 93.7% on average and for Grass carp 77.5%. The age and body weight of spawning schools vary with regions: In Pearl River drainage, smaller in size; earlier in maturity, generally, one year earlier than in Changjiang River drainage. and in Heilongjiang River 1—2 years later than in Changjiang River.

The brood amount of Silver carp, Bighead, Grass carp and Black carp is great and the measurement is usually done by way of sampling. It consists of the absolute and the relative brood amounts. The former refers to the total quantity of eggs borne by a fish in one spawning season, while the latter means the quantity of eggs per gram of female body weight in one spawning season. There is no relation-ship between eggs/g of ovary and fish size.

The eggs of Silver carp, Bighead, Grass carp, Black carp and Mud carp are separated and non-adhesive, The discharged eggs expand by absorbing the water through egg membrane and become plump, transparent and elastic. With a bigger specific gravity than water, they will sink to the bottom in still water and yet, they are semibuoyant in the currents, floating in the water till the fry come into being.

The incubation of eggs requires about the same temperature as needed in the spawning of brooders, between 18°c and 30°c with the optimum temperature of 22°—28°c. The speed of embryonic development is in positive proportion with water temperature.

Propagation of Common carp and Crucian carp

With the wide distribution, Common carp and Crucian carp do not require strict environmental conditions in the development of gonads and in the process of reproduction; therefore, their natural propagation can be easily found in the still or running waters of the South and the North.

Their spawning season generally comes at the end of March to early April in the southern part of China, April—May in the northern part of China, and June in the northeast part of China; however, it begins as early as in late December in Pearl River basin. Although the spawning season makes its start at different times, the demand for water temperature is identical, 18°c above for Common carp and 20°C above for Crucian carp. As the eggs of Common carp and Crucian carp are adhesive, the fundamental condition for spawning is the materials (aquatic plants or resemblances) for the eggs to adhere to.

The spawning is in full flourish from midnight till dawn. But it's possible that they could spawn in the whole day if environmental conditions are favourable. In estrus, two to three, males chase one female, and the male will hit their head against the female's belly time and again so as to make the female lying on its side towards the aquatic plants. In a minute, the female spawns while the male discharges milt. The action of spawning repeats so many times, that the matured eggs could be discharged in large amounts. And then the eggs could stick to the aquatic plants in water.

The age of individual fish's sexual maturity and its body size vary with different environments and climates. In Changjiang and Yellow River basins, Common carp and Crucian carp generally reach maturity at the age of 2 full years. In northeastern part of China, it takes more than 2 years with bigger size. The brood amount is related with fish size. The average brood amount of

Common carp, is 244,000 pieces for 3-year old fish of 44—48 cm in body length and 1.9—2.75 kg in body weight, 447,000 eggs for a 5-year old fish of 54 cm in body length and 3.5 kg in body weight. That of Crucian carp is 200,000—300,000 eggs for a fish of 0.5—1 kg in body weight. At 20°c, the fertilized egg of Common carp takes 101—104 hour to incubate into fry; at 25°c, 49—53 hours; at 30°c, 47—50 hours.

WATER LAYER FOR HABITAT

Silver carp, Bighead, Grass carp, Black carp, Common carp, Crucian carp and Mud carp live in different water layers. It's due to their different feeding habits. Silver carp and Bighead chiefly feed on planktons. They dwell mostly in the upper layer where there is abundant plankton. Grass carp and Wuchang fish like to search for their food in the upper and middle layers, or near the river bank, lake side or pond dikes, Grass carp, Common carp, Crucian carp and Mud carp, like to stay in the bottom layer, because they are all benthos-feeders.

Water Temperature

To a great extent, fish are poikilothermal or cold-blooded animals. The metabolism of fish is controlled by temperature. The appetite of the above mentioned fish is obviously diminishing when the temperature drops below 15°c and they stop eating below 5—7°c. As a whole, these fish except Mud carp have a wide-ranged adaptation to temperature. They can live in the temperature between 0.5°—38°c with the optimum of 25°— 32°c and yet, the optimum for food intake and growth is 25°—32°C (30°— 32°c for Mud carp). Silver carp, Bighead, Grass carp, Black carp and Wuchang fish begin to die when the temperature is below 0.5°c or above 40°c. Mud carp, T. nilotica and T. mossanbica will often freeze to death below 7°c, 8°c, 12°c respectively.

Water Quality

Silver carp, Bighead, Mud carp and Tilapia, feed on plankton and habituate themselves to the fertile water where there is plenty of plankton. Grass carp and Wuchang fish feed on grasses. Since Black carp feed on molluscs they'd like to stay in sheer water. Common carp and Crucian carp can live in all kinds of waters with their strong adaptation.

pH Value and Dissolved Oxygen

The four main Chinese carp require certain amount of pH value in the water, generally 7.5—8.5 pH, with light alkalinity as optimum. Living in the water of 6.0—10.0 pH for a long time will retard the growth of a fish. In addition, a fish must carry out normal gas exchange in water and demands a certain amount of dessolved oxygen in water. The higher the D.O, content the greater the feeding intensity. With D.O. above 4—5 mg/L, the feeding intensity of these fish is great, the growth fast, food conversion coefficient low. With D.O. below 2 mg/L, the fish lose their appetite, and below 1 mg/L, the fish will stop feeding and will gasp for air, and when the oxygen level drops further to below 0.5 mg/L, it will normally cause suffocation and death.

BIOLOGY OF ARTIFICIAL PROPAGATION

The artificial propagation means a series of processes from sexual maturity, ovulation, spawning upto incubation of fry under control by man. Gonads are also termed genital glands. The gonad of female fish is ovary, that of male fish testis. The eggs and sperms develop into maturity respectively in ovaries and testes when fish get to the age of sexual maturity. The gonads have cyclic variance in development, and in the process of it, they are controlled by internal endocrine and nervous systems and external ecological conditions. The two factors are associated and restricted each other.

STRUCTURE OF OVARY AND OVUM, STAGE DIVISION OF OVARY

The ovaries are in pairs, symmetric and sac-like. Their walls are formed by connective tissues and smooth muscles. The inner wall of ovary protrudes above septum, named as ovum-producing plate, where ova are turned out. Following maturity, the ova break follicular membranes and drop into the ovary cavity, the end of which adjoins a short oviduct with an opening to the exterior of body. There are blood vessels and nerve branches on the ovary tissue.

Just like other animals' eggs, fish's ova possess the common features of body cells, consisting of cytoplasm (ovum cytoplasm), cell nucleus (ovum nucleus) and cell membrane (egg membrane).

At the initial stage of development, the sphere-shaped egg cells only have nucleus and cytoplasm. In further development, the egg will accumulate egg yolk materials by and by. The amount of egg yolk is much larger than that of egg cytoplasm. Polarization appears in their distribution, thus the egg is called polarized yolk egg thereby.

The egg is topped by the so-called animal pole. Except for the yolk, most of the egg constituents such as cytoplasm and nucleus are concentrated near this pole. The yolk concentration increases towards the opposite, vegetal pole because of its larger granule and specific gravity.

After fertilization, the first mitotic division of the cell cytoplasm and nucleus will take place and embryonic development will begin. Egg yolk neither takes part in nor hinders the process of mitotic division. Egg yolk contains protein, fat, glycogen and vitamin etc., which are the nutritional materials for embryonic development. The nucleus consists of nuclear membrane, nucleole, nuclear fluid and chromosome. The nucleus is sphere-shaped but often leaf-shaped. Its function is to maintain genetic materials (DNA) and cell metabolism, passing the genetic material on to one generation after another.

The cell membrane is a membrane structure covering the exterior of egg cell, that can be divided into 3 kinds in accordance with the materials. Primary cell membrane or yolk membrane is made of cell's own plasm, with lots of radiant duct-like pores, so-called radiant belt, which are helpful for the absorption of nutrients and the discharge of metabolic wastes. Secondary cell membrane or chorion is secreted by follicle cells in ovary. The secondary cell membrane of fish is normally adhesive.

It is very common in adhesive eggs of Common carp, Crucian carp and Wuchang fish etc.. Tertiary cell membrane is composed of the secretion from glands in oviduct, such as gummy cell membranes of frog and cuttlefish. Fish do not have all the three above-mentioned membranes. Silver carp, Bighead, Grass carp and Black carp have just the primary cell membrane and Common carp, Crucian carp and Wuchang fish primary and secondary membranes.

THE DEVELOPMENT STAGES OF THE OVARY

The development stages of ovary maturity can be judged by visual observation or histologic survey. The development of fish gonad may be generally divided into six stages according to its appearance, colour, size, weight, and the distribution of blood vessels and the maturity of ova. Nowadays, the classification of the development stages of ovary adopted by each country is not identical. Five stages are admitted in India, Japan and the United Stages of America. Several countries prefer seven stages, whereas China does six. Set Silver carp as an example:

Stage I Ovary

By visual observation, the gonads are located at the lower part of air bladder, closely attached to the coelomic membrane, lineal in shape, transparent, flesh-white in colour and it's impossible to distinguish them into different sexes by the naked eye.

Tissue slice: cells tiny, diameter 12—22 um; nucleus rather big, occupying more than half of oocyte's diameter; very few nucleoli in the centre of the nucleus.

Stage II Ovary

By visual observation, ribbon-shaped, flesh-white in colour and semitransparent. It is impossible to distinguish ova one after another by the naked eye, and yet small eggs are visible when examined under a magnifyingglass; petaline-shaped when in fixation. It is possible to distinguish sexes. The maturity coefficient is about 1—2%.

Tissue slice: cell in multiangular shape or sphere-shape with a diameter of 90—300 um. A thin layer of ribbon-shaped follicles surrounds cell membrane surface. The nucleoli closely attach to the nuclear membrane .

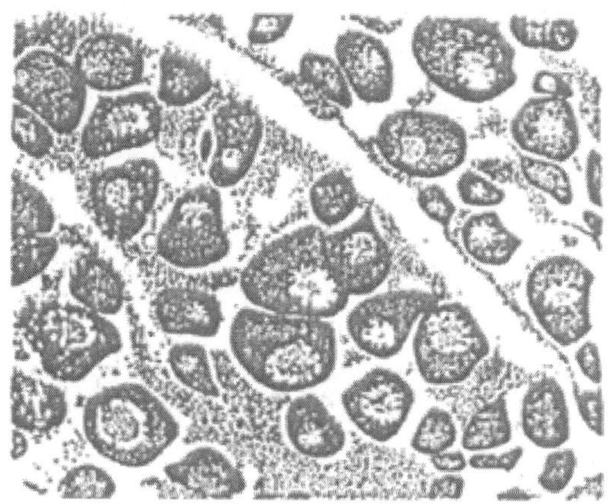

Fig. : Tissue Slice of Stage II Ovary of Silver Carp

Stage III Ovary

By visual observation, the capacity of the ovary has become conspicuously enlarged. It is due to the appearance of melanotic pigment so that it turns greenish grey in colour. Eggs are visible with the naked eye but not easily separable. The distribution of blood vessels is clear. The maturity coefficient of ovary is 3—6%.

Tissue slice: The follicular membrane in oocyte is bilayer. The egg yolk begins to form. One or two layers of vacuoles appear on the edge of the cell. The cell is 250—500um in diameter. The nucleus in the centre is irregular or oval-shaped. Greater part of nucleoli are distributed on the edge of nuclear membrane and small part scattered in the centre.

Stage IV Ovary

By visual observation, it is long and sac-lake, occupying 1/3—1/2 of coelomic cavity. Eggs are plump and greenish grey or light yellow in colour and easy to separate. Ovary is fully distributed with blood vessels. The maturity coefficient is 12—22%.

Tissue slice: egg yolk granules fill almost all the space outside nucleus with only a little cytoplasm spreading around nucleus and near egg membrane. The egg is enlarged to 800—1580 um in diameter. Nucleus edge looks like waves in shape with a few nucleoli inserted in the troughs and most of nucleoli moving to the centre.

This stage can be further separated into 3 stages in the light of occyte's diameter and nucleus location. Early stage IV: egg diameter about 800 um, nucleus in the centre; middle stageIV: egg diameter about 1000 um nucleus in the centre or inclined to move to animal pole; late stage IV: egg diameter about 1580 um, nucleus having moved to animal pole, which is so-called polarized phenomenon.

Lots of experiments and practices have proved that no matured eggs can be obtained by inducement as the oocytes of silver carp, Bighead, Grass carp and Black carp are at early stage IV. Only in case of middle and late stage IV when the nucleus is eccentric or polarized, matured eggs can be acquired, and then the artificial estrualization will succeed. These stages could last as long as 1—2 months, even 3 months for some cases, providing that proper ecological conditions for spawning are not available during that period of time or no artificial propagation is performed.

Stage V Ovary

By visual ovservation quantities of oocytes come out from follicular membranes into ovary cavity, and eggs are in a state of flowing. The ovary and the belly are very soft. The eggs would flow out from cloacal opening by way of a slight pressure upon belly.

Tissue slice: the yolk granules start to fuse into a lump. The cytoplasm and the nucleus have already moved to animal pole. The nucleoli concentre in the centre of the nucleus, whereas the nuclear membrane has dissolved through perforation. The nucleus looks transparent.

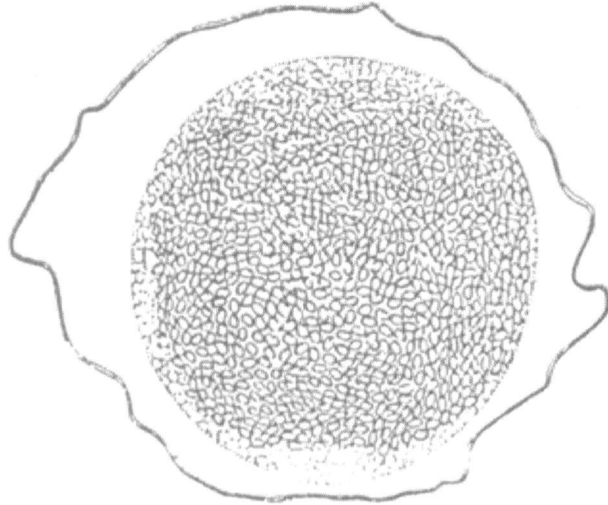

Fig. : Tissue Slice of Stage V Egg of Silver Carp

When oocytes proceed to their maturity, the epithelial cells of follicles secrete a sort of substance which can dissolve and absorb tissues between the follicular and the egg membranes; thus the eggs can easily be released out from follicles thereby. The eggs are in a state of flowing in the saclike. ovary.

This process is called ovulation. The eggs, which are in a state of flowing, are released from the fish body, this process is termed spawning. The oocytes proceed very fast from stage IV to maturity (stage V).

In nature, the process may be fulfilled in 20—40 hours after the rising of water in the rivers. Under artificial estrualization, it can generally be fulfilled within 10 to 20 hours or even in several hours. If the follicles discharged inmatured eggs ahead of time, the rate of fertilization would be affected. In another case, if the follicles could not release matured eggs in time, the eggs might be suffocated or degenerate by virtue of overripe and the rate of fertilization would be certainly affected too. Even if part of these eggs might become fertilized, the embryos could not develop normally. In other words, the success of either natural spawning or artificial insemination depends upon knowing exactly the maturity stage and the spawning time of a fish. If the matured eggs missed the chance of being released, the oocytes would degenerate and be absorbed.

Stage VI Ovary

By visual observation, the major portion of eggs has already been laid. There are still some phase-4 oocytes which have remained in the ovary. The ovary is slack and its size has been reduced conspicuously. Blood vessels have become enlarged with lump-shaped extravasated blood. Tissue slice: After ovulation, there are lots of follicular membranes and undischarged mature eggs in ovary. The undischarged eggs will soon degenerate and be absorbed, forming a semi-transparent and irregular structure, orangeyellow in colour. Besides there are still many interim oocytes in it.

THE STRUCTURE OF THE TESTES AND SPERM AND THE DEVELOPMENT STAGE OF THE TESTES

THE STRUCTURE OF THE TESTES

The testes are paired and tubular. They are situated on both sides of the air bladder, attached to the ceolomic wall. Mature testes are white in colour. There are a lot of ampullae arranged irregularly inside. The spaces between ampullae are full of connective tissues.

The ampullae are composed of many spores or seminal vesicle sacs (spermatangium spermatogonia). Spore sacs are separated by a thin layer of follicular cells. In one spore sac there is plenty of synchronously-developing germ cells, and in different spores sacs, germ cells vary in developmental stages. The centre of the ampullae forms a hollow cavity. After the formation of sperms, spore sacs dissolve and the sperms enter the cavity of the ampullae. The terminal end of the testis is connected to a short seminal duct with an opening to the exterior of the body.

The structure of Sperm

A sperm cell of Chinese carps consists of a head, a neck and a tail. The sperm head of Silver carp is almost spherical, 2.2—2.5 um in diameter, consisting of an apex and a nucleus.

The apex is situated at the front part of the head. It is also called penetrator for its function of penetrating into egg. The neck is very short and is situated between the head and the tail. The sperm neck of Silver carp is about 1.1 um in length. The tail, narrow and long, is many times more than the head in length. The sperm tail of Silver carp is about 35um in length. The tail is the metabolic centre and motor organ.

Mature sperms congregate in ampullae cavities because of the disintegration of spore sacs. Sperms will mix with the fluid secreted by the interstitial cells in sperms and the testis, forming so-called milt. The milt may be exuded or pressed out of body at the climax of brooders' estrus. 1 ml of Silver carp's milt holds approximately 48 million sperms. The total amount of milt could reach 30—40 ml.

The Development Stages of the Testis

Just as the ovary, the development of the testes may be divided into six stages according to its appearance and tissue slice observation.

Stage I testes: Testes are lineal in shape, transparent, attached closely to ceolomic wall. It is impossible to distinguish sexes. On the tissue slice, scattered spermatogonia, 16um in diameter may be observed. The nucleus is big and round, 9 um in diameter. Ampulli and seminal vesicles are still under formation; therefore, there is no clear fixed arrangement of sperm cells. Stage II testes: Testes are lace-like, transparent or non-transparent. Blood vessels are not clear. Histologic characteristics of this stage are multiplication of spermatogonia, formation of seminal vesicles which are arrayed in bundles. At this stage, ampulli are solid and separated by connective tissues.

Stage III testes: Testes are rod-shaped, pink or yellowish, elastic on surface with clear distribution of blood vessels. On the tissue slice, a hollow cavity may appear in the middle of solid ampullae with one or several layers of seminal vesicles on ampullar walls.

Stage IV testes: Testes are milky white in colour with a clear distribution of blood vessels on the surface. It is impossible to squeeze out milt at early stage, but possible at late stage. On the tissue slice, some larger primary spermatocytes, some smaller secondary spermatocytes and the smallest spermation can be observed. All these above-mentioned cells congregate on the walls of seminal vesicles with a small number of sperms.

Stage V testes: Testes are white in colour and full of milt. The milt will flow out through cloacal opening if a male fish's head is taken up and its belly slightly pressed. A large number of sperms and sperm cells in different stages could be seen inside ampulli on tissue slice.

Stage VI testes: The capacity of the testis has greatly decreased after milt exudation, yellowish-white or pink in colour on tissue slice. Only spermatogo nia, a little primary spermatocytes and connective tissues remain in seminal vesicles. After exudation, the testies of the male fish is turning back to stage III and will develop on.

SEXUAL CYCLE AND MATURITY AGE OF FISH

There is a certain sexual cycle in the development and maturity of the gonads of fish. When fish reach the maturity age and have their first spawning or milt discharging, their gonads will develop cyclically on without ceas in the light of season changes which is termed sexual cycle. In pond culture, the sexual cycle of Silver carp, Bighead, Grass carp, Black carp and Mud carp are basically the same. But, in nature, they spawn once a year, so the sexual cycle is one year. In the southern part of China, the climate is rather warm. Through intensive culture, pond-reared spawners after spawning in spring can again reach maturity in the same year and be induced to spawn 2–3 times a year.

The sexual maturity and the gonad development of fish are tightly associated with environmental conditions that is water temperature, feeds, oxygen etc. Accordingly, there is a variance of fish sexual cycle in different regions. Even in the same region, the gonad development could not be totally the same owing to various conditions.

Cyclic Variance of Ovary Development

(gonad development of pond-reared Silver carp in Zhejiang Province) In winter (November—next January), the ovary is at stages I—III with average maturity coefficient 5—6%. Huge amount of lipid in fish body occupies about 3.5% of the body weight. Eggs of old generation are at the stage of degeneration and absorption, some of them have already been absorbed. The new oocytes begin to accumulate egg yolk.

In summer (May—July), the ovary grows fast with obvious increase of weight, having passed through stage IV———maturity stage with maturity coefficient 17—20%. The lipid content of fish body reduce sharply; the belly is soft and expanded. If water temperature is favourable, inducement may be conducted.

In autumn (August—October), already after ovulation, the ovary is in atrophy with the maturity coefficient around 10%. The majority of those ovaries that have not ovulated show apparent degeneration, returning quickly from stage IV to stage II. In winter, ovaries will develop again from stage II or stage III. The nutritive materials such as lipid will be gradually accumulated for the next sexual cycle. As Mud carp are subtropical fish, the gonad development requires higher temperature. The gonad develops slowly in winter when ovary is at stage II. Upto next April (water temperature 26°C in the southern part of China), the gonad develops quickly into stage IV.

Cyclic Variance of Testis Development

The development of male fish's testis is earlier than that of female's ovary. Generally speaking, it has already reached stage III in water and could exude a little milt under pressure in early spring and before reproduction season. In autumn, after milt-discharging, the sperms remaining in testis will be absorbed through degeneration. The spermatocytes of a new generation begin to grow. The testis takes another new sexual cycle.

Maturity Age

Under different geographical and ecological conditions, the maturity age of the same species is widely different. The maturity age of Silver carp, Bighead and Grass carp in the southern part of China is 1—2 years earlier than that in the northern part of China both for the male and female. . And yet, even in the same region, it differs more or less under variant ecological conditions, for instance, in Jiangsu and Zhejiang Provinces, the maturing age for female Silver carp is 4 years, for Grass carp and Bighead 5 years, Black carp 7 years. If ecological conditions are suitable, the maturing age could be one year in advance. The males generally mature 1 year earlier than the females.

FECUNDITY

It refers to the number of mature eggs of a female fish before spawning or individual fecundity. The assessment of the fecundity of fish is based on maturing age, sexual cycle, brood amount, effective production of eggs and the survival rate of fry, etc.

Emphasis here is laid on the brood amount and spawning amount of pondreared Silver carp, Bighead, Grass carp and Mud carp.

Production of eggs: Under artificial propagation, the average egg production of Silver carp, Bighead, Grass carp and Black carp is 50 pieces per gram of body weight. The highest is 100 pieces/g. The average amount of Mud carp is a bit higher because the eggs of Mud carp are much smaller. .

RELATIONSHIP BETWEEN ENDOCRINE SYSTEM AND GONAD DEVELOPMENT

Just like other vertebrates, all physiological activities inside the body will be regulated and controlled principally by nervous system and endocrine system, of which the pituitary gland or hypophysis, gonad and thyroid are closely associated with the gonad development of fish.

STRUCTURE AND FUNCTION OF HYPOPHYSIS

The hypophysis of fish is located below on the ventral side of the thalamencephalon and attached to the hypothalamus. It's divided into two parts: the neuro hypophysis and the adenohypophysis. the neuro hypophysis is directly connected to hypothalamus with its nervous fibres and blood vessels planted deep into the adenohypophysis. The adenohypophysis could be divided into anterior lobe (pro-adenohypophysis), the transitional lobe (meso-adenohypophysis) and the posterior lobe (mota-adenohypophysis). The anterior lobe is situated nearest to the thalamencephalon where there is a little distribution of nervous branches and blood vessels. The transitional lobe is situated at the lower front of the anterior lobe and the posterior lobe is situated at the lower front of the transitional lobe.

The meso-adenohypophysis of the hypophysis of fish contains a sort of basophils which can secrete sexual hormone, *i.e.* FSH (Follicle stimulating Hormone). These agents can stimulate the growth, development, maturity and ovulation of eggs. They can either promote the synthesis secretion or estrogen or the formation of sperm and the secretion of antrogen.

The sex stimulating hormone of the hypophysis of fish varies with the ages and seasons. As a rule, the secreted amount of hormone by mature fish is great er than by immature fish and higher before spawning than after spawning. Not only the hypophysis of the same species can be used for estrualization but also that of the same genus can be basically applied to one another and the hypophysis of most of the same family can be adopted reciprocally. For example Common carp and Silver carp do not belong to the same genus but do to the same family. If the hypophysis of Common carp is used to induce Silver carp for spawning, it is effective and vice versa.

Gonad: The gonads of fish produce germ cells —sperm cells and eggs and secrete sexual hormone as well. The male sex hormone secreted by the testes is called antrogen. The female sex hormone secreted by the ovary is called estrogen. The hormone causes the development of subsidiary sexual organs and secondary sexual character and the appearance of sexual behaviour. Other endocrine glands affect the development of gonads directly or indirectly, *e.g.* the thyroid gland secretes thyroxine which can provide the indispensable energy for spawning under low temperature; adrenal cortex (internal tissue) secretes adrenal cortex hormone which can regulate the metabolism of carbonhydrates and also function in the control of salt and water balance.

FUNCTION OF NERVOUS SYSTEM IN PROPAGATION

Propagation of fish is a rather complicated process, which demands that a living organism have a continuous balance between the internal make-up and the external ecological conditions, only by which can the organism survive. The realization of this balance relies upon the coordination of nervous system and body fluid regulatory system and their mutual restriction.

The whole process of the gonad development of fish is to a great extent, con trolled by the hypophysis, whereas the secretion of the hypophysis is in turn controlled by the external factors through nervous system.

In nature, the spawning of Silver carp, Bighead, Grass carp and Black carp from the relative stage IV to spawning stage V is fundamentally controlled by the external ecological conditions during the natural reproduction process It is easy to understand by the following block diagram.

It shows that when certain ecological conditions— light intensity, water temperature, running water and opposite sex stimulate the external sense organs—skin, lateral, senses of sight & hearing, the nerves of these organs will give impulses immediately, which will be conveyed into the central nervous system and they will in turn, urge hypothalamus to release a hormone or polypeptide, *i.e.* LRH. This hormone stimulates through the portal vein or the hypophysis, the basophilous cells in the hypophysis to release (LH) leutinizing hormone and follicle stimulating hormone (FSH), then these hormones flow into the gonad through blood circulation, promoting the growth and development of gonads. Meanwhile, gonads also secrete a sort of sexual hormone, which can affect the hypothalamus and the hypophysis in return making the spawners and the milters perform sexual activities of chasing each other an estrus, natural courtship, spawning or releasing milt.

INFLUENCE OF ECOLOGICAL CONDITIONS ON GONAD DEVELOPMENT

Owing to their continuous adaptation to external environments over their long period of evolution, fish have formed their respective demands for different ecological conditions. Fish will demonstrate a fine growth and development if ecological conditions are favourable. If the relevant ecological conditions are lacking, the growth and development will, to some extent, be restricted; if the conditions are excessively unfavourable, fish will die.

The principle ecological requirements are nutritions, temperature, current, dissolved oxygen and light intensity etc. These conditions lay a constant and comprehensice effect on the growth of fish and the development of gonads.

Nutrition

Only under a rational nutritional condition can the gonads of fish develop well; hence, at the early developmental stage of the ovary, the maturity coefficient is generally 5—6%. The nutrient substance, the egg yolk starts to accumulate, therefore, fish obtain protein and fat from outside for the development of gonads and convert the excessive surplus energy to fat for storage. As a result, fish are quite fat at that time. In late spring *i.e.* late April or early May with the rising of temperature, the ovary grows quickly the maturity coefficient rapidly increases to 12—20%. The fish have accumulated large amount of nutrient substance for the egg yolk and continue to absorb nutrition from outside for maintaining normal living activities, then they transform the stored nutritional fat into egg yolk nutrients for the growing & development of gonads.

The facts above-mentioned show that if the broodfish have good rearing conditions and are able to accumulate sufficient nutrient substance in autumn after spawning season, furthermore, the nutrition level and good conditions are maintained in spring, the gonad of fish would mature earlier and the broodamount would be great; if not, the development of the gonads of fish might be restricted or even the gonads could not reach maturity.

Among the nutrients such as protein, fat, carbohydrates, vitamin and minerals, the most needed for the development of the gonads is vitamin E. On the basis of practices, better results come from the supplies of feeds full of vitamin E in early spring, like wheat sprouts, rice sprouts, lettuce leaves and Lactuca Indica etc.

However, the blind offering of abundant feeds with a neglect of other living conditions will make the fish rather fat, while the development of the gonads will be restrained; therefore, the nutrition is not the only factor. It must be combined with other conditions so that the gonads can have a better development and reach maturity.

Water Temperature

Water temperature is a significant factor affecting the metabolic rate of fish. The maturing age of fish is also closely with water temperature. It also affects the developmental rate of the gonads of fish.

Remarks:

1. The growth period is counted when the monthly average water temperature is above 15°C.
2. Total heat during the growth period, the formula is as follows: Total heat = average water temperature during the growth period × days
3. The formula : Total heat during the maturity period = total heat during the growth period × maturity age

Owing to diferrent water temperatures and the growth periods in South China and North China, Silver carp have different maturity ages; anyhow, the total heat needed for maturity is basically identical, about 18,000–20,000 degree/day. This shows a positive relationship between the gonad developmental rate and the water temperature (heat). To raise the water temperature is an effective measure to mature the gonad and to get early estrualization during the broodfish culture period in the North.

Running Water

The letting-in of fresh water into broodfish rearing ponds at definite or indefinite time can keep water quality good which makes fish comfortable and is benefitial to the growth of fish. At the same time it can regulate the composition of the natural feed, in order to make it more reasonable and to raise the nutritional level of the broodfish. The stimulation of running water has a marvelous effects on the development of the gonads, especially when the germ cells develop to stage IV. It will accelerate the process of metabolism and the process of transformation from stored nutrients to the gonads. Furthermore, according to experiments, it may stimulate hypothamamus to synthesize and release large amounts of LRH, which will further stimulate hypophysis to release sex stimulating hormone, then the spawners are induced to estrualize and lay eggs. Slight running water all year round is very effective to Silver carp. Slight running water all year round and letting-in of fresh water before spawning have the same effects on Grass carp, but no running water is not so good for spawning of Silver carp and Grass carp and fertilization of eggs.

DISSOLVED OXYGEN

Oxygen is one of the most elementary living factors for fish. It guarantees the metabolism of fish through their respiration. When the dissolved oxygen content is 2 mg/L in ponds, the normal physioligical activities are sharply affected with fish gasping for air. The excessive consumption of energy inside fish body will harm the development and maturity of the gonads of spawners. In such case, most of the induced brood fish fail to spawn properly. Especially as spring comes, the demand for oxygen is becoming urgent, usually above 4—5 mg D.O./L. If the water is clear with rich D.O.C., the brooders spawn normally. In brood fish culture, due attention must be paid to manuring, feeding amount, stocking density and the timely letting-in of fresh water and continuous improvement of fish living conditions.

Besides the aforementioned ecological factors—nutrients, water temperature, dissolved oxygen, which promote the development and maturity of the gonads of fish, there are also light intensity, salanity, and the presence of opposite sex etc. Each of these is part of comprehensive factors which control the progress of the sexual cycle of fish.

THE ADVANTAGES OF INTEGRATED FISH FARMING

In developed countries, the more their industry develops, the more intensive their farming is, and the larger the scale of chicken and pig raising factory farms is, the more the wastes accumulate, *e.g.* in Japan, only the excreta of cattle, pig and chicken reach more than 70 million metric tons per annum. If this great amount of excreta is not disposed of, it must pollute the environment. The annual output of animal manures is considerably high — 10.4–11.8 tons per black-and-white cow; 4.1–4.7 tons per pig; 0.04–0.07 tons per duck. The decomposition of so much excreta produces gaseous ammonia, etc. The odour pervades everywhere. As a result, the water, the land and the air are all polluted and it will jeopardize the people's health.

In fact, the manures of livestock and poultry are good organic fertilizers for fish farming. About 40–50 kg of organic manures can be converted into one kg of fresh fish. If the farm combines fish farming with mulberry cultivation, sericulture and silk extraction from the cocoons, the pupae will be a fine feed for fish while the worm faeces and waste water obtained at the processing factory are used as pond fertilizers. In an integrated fish farm, the manures of livestock and poultry and the dregs, lees and waste water obtained from starch processing and wine brewing are all used for fish farming. The pond silt can be used as fertilizers for terrestrial fodder crops which can in turn be used to raise livestock and poultry or directly for fish farming. Thus, it forms a recycling ecosystem, which utilizes various wastes and afterwards, an agricultural ecosystem can be set up without any wastes.

INCREASING THE FOOD SUPPLY FOR THE MANKIND

Short supply of food and protein is a serious problem that we are now facing. Therefore, merely using the pelleted feeds of grains and animal protein as fish feeds is not economical and often makes the food for the mankind more insufficient. If the grains such as wheat are used in fish culture, the output could be up to 4815–9750 kg/ha. But the average Food Conversion Rate of grain is 3, *i.e.* 3 kg of grains in dry weight could converted into 1 kg of fresh fish. This kind of fish culture is not suitable for the countries which want for food. However, the natural food organisms cultured in fish ponds by using organic manures could totally take the place of pelleted feeds or grains.

The quality of fish will not be changed by using animal manures as fertilizers in fish culture. The daily output could reach 15–32.25 kg/ha or even higher. E.g. Helei Fish Farm in Wuxi has not only built fish ponds with an area of 69.4 ha but also a dairy with 100 cows and pigsties with 1000 pigs and duck yards producing 10,000 eggs per day. It has become a subsidiary food production base, which is an integration of aquaculture industry and commerce. So far as aquaculture is concerned, they combine fish farming with livestock and poultry raising, however, putting fish farming first. In 1981, it supplied not only large amount of fertilizers to produce 600,000 kg of fish for market but also supplied 490,000 kg of pork, etc.

The scope of integration in an integrated fish farm could be considerably wide. Apart from fish cultured in the water body, the water surface can be used for goose and duck raising, pond dikes for fruit tree and mulberry cultivation or for setting up pig-sties, slopes for fodder crops. The products from an integrated fish farm are not only fish but also meat, milk, eggs, fruit, vegetables, etc. It is apparent that integrated fish farming can fully utilize the water body, the water surface, the land and the pond silt, etc. to increase the food supply for people.

More Job Offer

In unitary fish farm, the labour force cannot be reasonably employed while the integrated fish farm may offer more jobs, for example, in Helei Fish Farm, the occupations except fish culture offer 149 jobs, among which there are 48 persons for duck raising, 19 for cow farming, 14 for pig raising.

Increasing Economic Benefits

Nowadays, the problem that the aquaculture in developed countries is facing is that the cost of pelleted feeds is too high. It is related to energy crisis and protein shortage also. The integrated fish farm produces feeds and fertilizers for itself, savrer energy and reducing. expenditure considerably, *e.g.* in 1981, Helei Fish Farm produced 5.5 million kg of pig manure, 1.85 million kg of cow dung, 1 million kg of duck manure, 9.5 million kg of waste water from silk extracting workshop. The total amount of organic fertilizers was 17.65 million kg. Outflow of animal excreta by gravity could reduce the production cost by about 97 year each mu of fish pond. In 1976, Helei Fish Farm began to put duck, cow and pig raising into integrated fish farming system, food processing industry in 1979 and commerce in 1980. During the period of 1977–1981, the yield of fresh fish increased from 115,000 kg to 600,000 kg (the same area) while the yields of livestock and poultry increased from 135,000 kg to 490,000 kg. In 1981, average income per capita increased by 120. over the year of the establishment of the fish farm in 1966 . The annual net fish production of integrated fish farms, only by using organic manures.

CHARACTERISTICS OF INTEGRATED FISH FARMING

The integrated fish farming practices have developed into a complicated structural network in line with the local conditions. China is a vast country with a large population and varied natural environments. The agricultural structure and the economic conditions of each locality are also different.

Therefore, various integrated fish farming systems in line with the local conditions have developed into a complicated structural network. The Pearl River Delta is located to the south of the Tropic of Cancer. Its annual solar irradiation is 110 kcal/cm^2 but averaging 60 Kcal/cm^2. Average temperature is about 22°C; 2–3 days of frost a year; more rain and high temperature in summer with maximum temperature of about 37°C and relative humidity 76–85%; annual sunshine time between 2000 and 2500 hours. Such geographic and climatic conditions are very conducive to the cultivation of mulberry trees, sericulture and fish farming. Hence, the farmers in the Pearl River Delta through production practices have been able to take advantage of local natural resources by integrating mulberry cultivation and sericulture with fish farming, which leads to the establishment of a complete, scientific "mulberry plot-fish pond" man-made ecosystem.

Even in the same geographical zone, the items of integration by various production units are different. Helei Fish Farm, Xinan Fish Farm, Helei First Fishery Brigade, Wangzhuang Fish Farm as well as the Municipal Fish Culture Farm in Wuxi practise different items of integration. . Liutan village has recently incorporated agriculture, sideline occupations, aquaculture and commerce into its integrated system. The integrated fish farm was established in 1980. The farm site was earlier a water-logged paddy field of 17.33 ha, subject to annual flooding and with low yield of agricultural crops. Based on the topography of the low land, the farmers restructured the land into fish ponds of 10.67 ha. in the winter of 1979 and the spring of 1980.

The farm consists of 15 grow-out ponds (8 ha), 11 fingerling ponds (2.67 ha). Apart from pig raising, green fodder planting is the major item of the integrated management. It has 1.33 ha of pond dykes for the cultivation of English rye grass (Lolium pereme), sow thistle (Lactuca indica), etc. They utilized water cane shoots field to cultivate duckweed with average yield of 37,500–45,000 kg/ha.

With sufficient supply of green fodder, Grass carp and Wuchang fish (Megalobrama spp.) become the dominant cultured species, with the yield of 5250 kg/ha, that is 45% of the total pond fish production. It indicates that the maximal efficiency of the integrated fish farming system can be obtained only through maximal utilization of natural conditions and the agricultural characteristics of the regions concerned.

The socio-economic conditions should be taken into consideration in developing integrated fish farming. The fact is that the diversified economy develops in the harmonious interaction among the socioeconomic conditions, agricultural production, and natural conditions on the basis of regional climatical differences.

Since the development of integrated fish farming is site-specific and the conditions of each site are different and complicated either from the macro or micro view points, the linkage between the trades of integrated fish farms in China become a complex matrix. Simple modele involving mono-integration of fish-cum-animal husbandry, fish-cum-poultry or fish-cum-crops are now getting fewer and fewer in China with the exception of small-scale individual farms.

Utilising Wastes by Various Ways of Recycling

The integrated system of fish farming is rather complicates involving various forms of integrations and methods in utilization of organic wastes. The animal excreta alone can be effectively utilized through employing various methods and techniques.

1. Fresh animal manure can be applied directly to the fish ponds. pigsties, poultry coops and pens for ducks and geese can be constructed on the dikes or above the ponds. Fresh manure thus enter the ponds directly, avoiding energy losses due to processing of manure and transportation. The feedstuff of livestock are not fully digested and then can be directly utilized by fish. Therefore, the number of animals should be compatible with unit water surface.

2. The residues and liquid part after anaerobic fermentation to produce biogas are used for fertilizing fish ponds.
   Compost after aerobic fermentation can also be used for fish culture.
3. Animal manure can be used indirectly through one or two trophic levels in a food chain such as growing fodder crop to feed herbivorous fish, producing earthworms or other animal-feedstuff for carnivorous fish directly or used as part of the composition of pelleted feeds.
4. Poultry manure can be used to feed pigs and pig manure can in turn be used as fertilizers for fish pond.

APPLICATION OF NEW TECHNOLOGY STIMULATED AND IMPROVED THE PRODUCTION OF INTEGRATED FISH FARMING

The prosperity of agricultural economy and the application of new technology stimulated and improved the production of integrated fish farming, *e.g.* the use of aerators and the mechanization, to certain extent, raised production efficiency of the classical mode of Chinese fish culture. Although integrated fish farming has long been practised in China and rich experience has been gained through years of improvement and considerably high yields have been achieved, the scientific basis for some of the existing techniques are still awaiting clarification.

In order to raise the efficiency of integrated fish farming considerably, the biological basis of integrated fish farming must be studied, finding out the rule and developing aquaculture technology. In theory, two ecosystems must be clarified: First, the integrated fish-livestock-crop land-and-water ecosytem research should be carried out on its structure and functions so as to set up an optimal and harmoneous ecosystem on a perfect structure and with sound functions; second, the pond ecosystem.

The structural research concerns the biological interaction among fish, livestock and crops and their matched proportion. Material cycle and energy flow should be measured to elucidate the biological relationship and exact proportion of quantity among these production links. So far as the land-and-water structural ecosystem of integration is concerned, crops are producers; livestock and fish are consumers; aquatic organisms and soil organisms are decomposers. It is a rather complete ecosystem. In this system, benign circulation demands that there be no wastes in any link of this structure and the production be not hindered by lack of sufficient energy. For this reason, it needs a thorough research on fish pond ecosystem. Manureloaded pond is a semi-closed man-made ecosystem. Being put into fish ponds, animal manures begin to enter into the process of decomposition by bacteria. It's a complex process of pond dynamics in which animal manures will be converted into fish protein. It involves many factors such as physical or chemical parameters or biological factors in food chains.

In order to investigate the rule of variation and intricate relationship between biology and non-biology, the research needs cooperation and common efforts of different scientists and specialists from multidisciplines such as aquaculture, ecology, botany, microbiology and chemistry.

Integrated fish farming is a low energy consumptive but high efficient aquaculture system. It's a right way to develop fresh water fish culture and also it's a strategic measure of developing Chinese aquaculture. Integrated fish farming will play a promising role in solving the problem of "hard to get fish".

INTEGRATED MANAGEMENT OF FISH AND CROP FARMING

Fish-cum-crop integration is the most ancient and popular pattern of integrated fish farming.

Why can we combine fish farming with crop cultivation? It's due to the demand of fish feeds and the excessive pond silt. On one hand, abundant silt deteriorates the pond water and yet on the other hand, it is one of high quality manures for agriculture to plant fodder crops which can in turn be used as feeds for fish. Therefore, pond silt is a link between fish and crops in this integration.

Formation and Function of Pond Silt

Large amount of feeds and manures are put into fish ponds year after year. There is a considerable amount of residues and manures which often sink to the bottom. Moreover, the excrements of fish and aquatic animals, the corpse of the aquatic, and alluvial soil incessantly descend to the bottom too. The sedimentary organic material decomposed by bacteria forms a great deal of humus, which in turn combines with sludge on the surface of the pond bottom to form silt.

The thickness of silt varies greatly with many factors. Even in the same pond, the thickness in different locations is different. The mean thickness formed in a high-yielding earthen pond with target yield of 500 kg/mu and with no slope protection is about 10–20 cm, which is about 50–95 m^3/mu/year or 100–190 tons in wet weight.

The proper amount of silt is beneficial to fertilize the pond water whereas the excessive amount is not so good to fish. Microbes propagate rapidly when large amount of feeds and manures are applied. The pH value will decline, the BOD will increase, and gases such as nitrites NH_3, H_2S, CH_4, PH_3, etc. and surplus nitrites will do harm to fish. The median tolerated limit (TLm) value of Silver carp and Bighead fry in 24 hours at the temperature of 25°C is 0.91 mg/l and 0.46 mg/l respectively. Grass carp is more susceptible . Surplus nitrites will easily induce hemorrhagic septicemia of fish. Silt contains a lot of ichthyopathogen parasites and other harmful creatures. The more the thickness of silt, the more the deterioration of the pond water. It directly affects fish yields and thus, the excessive silt should be removed after the pond is drained. On the other side of the coin, silt is a manure of high quality which contains several nutritive elements. In Pearl River Delta and Taihu Lake basin, the pond silt is often removed 3 to 6 times every year.

The nitrogen content of pond silt/mu/year is equal to 481 kg of ammonium sulphate. Furthermore, as pond silt is quite complete in nutrition, it contains not only nitrogen but also phosphorus and potassium. Generally, T! of silt is removed per mu per year and that is equal to 1.2 tons of N.P.K. fertilizers. The quick-acting component of silt is equal to 119 kg of fertilizers, which can also serve as additional manure besides the basal one. Silt could increase the thickness of cultivation layer, improve the soil particle structure and strengthen its ability to absorb ions of N.P.K. and the ability to keep the water. It is beneficial to the prevention of fish diseases. Since it is also a slow-acting fertilizer which is beneficial to late crops. In terms of mass's experience, the quick-acting component of 100 kg of silt in dry weight can increase 1 kg of rice. If we only use silt as fertilizer, 100 kg of silt in dry weight can produce 10 kg of rye grass. Provided that 50 m^3 of silt is removed from each mu of fish ponds, all the silt can be used to cultivate 6 mu of rice, and then it could increase about 500 kg of rice; if it is used to cultivate rye grass, the output can reach over 6000 kg/mu. The yield can be increased by 2.5 times. From investigation data, the yield of per-unit paddy field by using 10–15 tons of silt composted with green grasses is near to the one by using 5 tons of animal manures composted. The cost of digging silt in China is cheaper than the one of purchasing animal manures from outside the fish farm.

The Demand of Feeds in Ffish Farming

The demand of both commercial feeds and natural food organisms is great in fish farming alone. The feeds and foods sometimes are hard to get and the great amount of supply also affects the balance of the market and ecology. Besides, the cost of transportation and energy consumption is surprising. The fish pond with the target net yield of 250 kg including 100 kg of herbivorous fish, 100 kg of planktoneater, 50 kg of omnivorous fish needs 1500 kg of aquatic grass, 1000 kg of vegetables, 150 kg of grains apart from manures. In order to meet the demand of feeds and foods in fish farming and to reduce the cost, it's necessary to combine fish farming with crop cultivation.

The Feasibility of Fish-cum-crop Integration

Aquaculture can provide large amount of silt and fertile water for agriculture. There still exists a potentiality of land on fish farms. E.g. the mean pond dyke is three meters wide, the gradient of a slope is 1:1.5–3. The mean area of a fish pond is 10 mu. The arable area of pond dyke and slope and the area of the water surface is in the ratio of one to five. The arable land might be greater before May with shallow water in the pond. With extra 0.3 mu of forage field attached to one mu of fish pond plus all the areas available, the average forage field for one mu of fish pond could amount to 0.5 mu or even higher which can provide fodder crops for herbivorous fish and proportionable filter-feeding fish in one-mu fish pond with net target yield of 400 kg. What's more, it's possible to plant aquatic plants on scattered unused surface. In short, it is necessary and feasible to integrate fish farming with crop production so as to fully utilize pond silt, arable land and water surface. As a result of that, the needs of fish feeds could be satisfied wholly or partially.

CROP VARIETY AND VARIOUS CULTIVATION SYSTEMS

According to crop variety and various cultivation systems, fishcum-crop integration can be divided into four patterns: fish farming integrated with terrestrial crop production; cultivating crops in ponds first, fish farming next; fish farming combined with aquatic plant cultivation and dyke-pond system.

Fish-cum-terrestrial Crops Integration

All or most part of crops planted in the fodder crop field and corner plots on pond dykes and slopes are used as green fodder for fish and as fertilizers for ponds. This is the most popular pattern in fish-cum-crop integration.

Crop Variety

Choose crop variety which are palatable to fish, rich in nutrition, strong in resistance to diseases, easy to manage and has features such as well-developed root, which protects the slope, etc. If it serves as straw manure, it should be easy to decompose. Besides, the average yield of some leguminous plants such as Trifolium repens, T.pratense, Medicago sativa, Astragaluo sinicus can reach 5000–7500 kg/mu. These grasses serve as both feeds and fertilizers. The average yield of some gramineous plants such as Pennisetum purpureum, Phalarosarundinacea, Pennisetum alopecuroidescross, P. purpureum can reach over 10,000 kg/mu. The seeds and young crops of other grains such as barley, wheat, maize and rice are also palatable foods of fish. The tender and juicy vines and leaves of sweet potatoes Ipomoea batatas and squash Cucurbita are pulverized as fish feeds for Grass carp in Wuxi area. The tuber and squash, etc. are cooked to feed feed-eaters.

COLLOCATION OF FISH FARMING AND CROP CULTIVATION

The ingestive variation of fish is caused by the growth of fish and environmental conditions. Among the environmental conditions, the seasonal change of water temperature is the main factor. The production period, therefore, has to be collocated with this variation of fish ingestion so as to synchronize the daily production of fodder grass with the daily ingestion of fish. The period of the lowest ingestion amount of cultivated fish in Changjiang Drainage is between Feb. and Mar. and between Nov. and Dec. in a year. The peak is between Jun. and Sept. accounting for 50% of the annual total amount. The production mode of English Ryegrass and Sudan grass nowadays coincides with the demand of fish feeds.

Ryegrass is sowed in Sept. every year and transplanted in Oct. and mowed in Dec. in Changjiang Drainage. It reaches the peak production between Apr. and May. The integrated fish farm can supply itself sufficiently by the production of Ryegrass. Sudan grass is sowed in mid Apr. and mowed for the first time when it grows to 50 cm high just before the withering of rye grass. It reaches the peak production between June and Sept. The daily output is about 100 kg/mu. Based on the experiments the annual production of 2 grasses can reach about 15,000 kg/mu if it's well managed. The other ways of collocation are as follows:

A. To intercrop gramineous grasses: Ryegrass, Sudan grass and leguminous fodder grasses.
   The unit area output of two families of fodder grass can be increased and the quality can be improved.
B. To adopt ensiling method if there is a surplus of ryegrass or to pelletize ryegrass with other materials, so as to be supplied during crop change in June as supplements.
C. To adopt sowing and transplanting by stages and to mow the grass in turn.

In Changjiang Drainage the seeds of Sudan grass are sowed by stages in April ahead of the usual time and can be mowed in mid May. In order to maintain the fertility of the land, pond silt must be fully utilized with some additional organic fertilizers apart from planting leguminous grass, and at last, the grasses should be harvested with stubble remaining in the field.

The Ratio of Water Surface and Crop Field

The ratio of water surface and crop fields means the total area of crop fields collocated with one mu of water surface of fish farming. The following factors must be taken into consideration:

a. Source of feeds and fertilizers and their prices.
b. The abundance of land.
c. The yield & the proportion of various spp.

Stocking Models of Fish in Fish-cum-crop Integration

The main or only source of feeds and fertilizers is the pasture grasses cultivated on the farm. The spp. stocked should be grass carp and Wuchang fish megalobrama spp. as major spp. which occupy 60–70%. Silver carp & Bighead as minor spp. which occupy about 20–30% assorted with a few of omnivorous fish which occupy about 10%. The omnivorous fish can not only utilize residues and detritus but also can clean the fish pond for herbivorous fish but the ratio should be limited within 10%.

In order to get natural food organism ready for Silver carp and Bighead after stocking, it is necessary to mow the grass and make a compost to fertilize the pond water or stock Silver carp and Bighead half a month later after stocking of Grass carp. If the farm produces more compost, the proportion of omnivorous fish may be increased a little.

The use of pond silt in fish-cum-crop integration There are 2 methods to use pond silt in accordance with crop strains and cropping system:

a. Pond silt is directly used as base manure for Ryegrass, etc. The silt should be harrowed and smoothed after it dries a little and then the seeds can be sowed. If it's a transplantation, there's no need to harrow. The seedlings could be interpolated, economizing on much labour and time. If the silt is used as additional manure, it should be topdressed to the roots of the plant. In summer, silt will be bailed with water to the fields. This method is beneficial for the silt to release or diffuse nutrient elements and detritus. It also benefits increasing of dissolved oxygen in the bottom layer of water.
b. Making compost with pond silt and grass
   Usually, there's a surplus of pond silt in winter whereas there's a shortage of pond silt in summer. Therefore dig out a pit, 5 meters square with a depth of 1 m near the crop field and then put in pond silt and grass (straw manure and stable matted grass are better) to make compost and cover the pit with mud. After fermentation, the effectiveness of manure is increased. The compost is usually used as base manure, sometimes mixed with water as topdressing.
   It's safer to use pond silt as manure. The applying amount is not necessarily limited, with an average rate of 5–15 tons/mu or even a thickness of 35 cm on the soil.

THE EFFECTS OF "ROTATION" OF FISH AND GRASS

The system of rotation of fish and grass can fully utilize the productivity of fish ponds to provide both green fodder crops for fish and green manure crops for fish ponds to propagate large amount of natural food organisms. According to the practices of Suzhou Municipal Fish Farm, the phytoplankton can reach 3×10^8 ind/1 at the peak time. The growth of summerlings can reach 1.8–2.5 mm/day in the initial month. If the yield of barnyard grass in a pond is 5000kg, 40kg of grass can be transformed into 1 kg of fingerlings averagely. It can save 0.5–0.8 kg of marketable feeds and reduce 10% of the cost.

In this system, crops utilize the nutrients of pond silt, whereas crop nutrients, which return back to the pond after submergence, not only improve the water quality but also increase the soil fertility. Practices prove that the fish yield increases after submergence of crops, *e.g.* Suzhou Municipal Fish Farm has adopted this method to nurture fingerlings since 1979. In 1982, the average yield increased from 150 kg/mu to 391 kg/mu, increasing by 1.6 times.

Fish Farming Integrated with Aquatic Plant Culture

In a network of rivers like lower reaches of Changjiang River, a fish farm is often near lakes, rivers or water-logging area or in the vicinity of inlet and outlet of irrigation canal. These water bodies are rich in nutrients, esp. effluents from cities and fish farms. In order to utilize these water resources, fish farmers in the southern part of China often culture aquatic plants in these water bodies. The principal aquatic plants are water hyacinth Eichhornia crassipes, water lettuce Pistia stratistes and water peanut or alligator weed Alternanthera philoxeroides, so-called three Ap's; the secondary are duckweed such as (Spirodela polyrhiza) and Wolffia arrhiza and Lemuna minor.

The Ways of Utilization of Aquatic Macrophytes

The output of aquatic macrophytes is the highest among the green fodder crops for fish. The yield could be about 15 ton/mu, even above 25 ton/mu. The aquatic macrophytes grow too fast and it causes a lot of trouble in the Tropics and sub-Tropics. However, the nutrients of aquatic macrophytes are higher. Water hyacinth is called "King of Aquatic Plants". So far as protein produced from per unit area is concerned, it is 6–10 times that of soy bean. Aquatic macrophytes are easy to manage with less labour and lower cost. It's said that one labour can manage 50 mu of 3 aquatic plants and can produce 13.1 ton of crude protein in half a year. The cost of one ton of three aquatic plants (including wages) is only about 1 yuan.

THE METHODS OF UTILIZATION

Three Ap's Directly Serve as Ffish Feeds

Three Ap's are processed in different ways to turn into different sized feeds for a variety of fish spp. in different size. To nurture fry, 3 Ap's should be mashed into grass paste, and the residues of leaves must be filtered out; then the paste can be sprinkled to the whole nursery pond. Table salt, 2—5% of the weight of the plant should be added to the water hyacinth paste to decrease the toxicity of saponin. The equivalent ratio of three Ap's and soy bean in fish farming is 17.5—25 : 1. It needs 5 kg of soy bean to nurture 10,000 ind of summerlings. We can use 87.5—125 kg of 3 Ap's instead. The cost of soy bean is 13–18 times that of three Ap's. Fry will grow fast with higher survivability if three Ap's are used as fish feeds. For example, Zhuang Aquaculture Brigade, Wuxian Country stocked 80,000 fry in 1977. After 16 days of nurturing, the transferred size was 3,33 cm, with 94.5% survivability from the pond fed with water hyacinth paste while the transferred size was 2.93 cm with 92.8% survivability from the pond fed with soybean milk.

To rear fingerlings three Ap's are esp. good for Silver carp and Bighead. 3 Ap's should be mashed into grass paste too but it's not necessary to remove the residues. To rear adult fish, with herbivorous fish of different spp. in different sizes as major spp. three Ap's are often pulverized by green fodder crops pulverizer and then fed to the fish. According to the experiments, about 45 kg of three Ap's could be converted into 1 kg of fish. The macrophytes per-mu can be converted into more than 400 kg of fish. If rice and wheat brans are used as feeds, it will need 1600 kg of brans to produce the same amount of fish. The cost of three Ap's is about 10% of that of brans, but three Ap's are less effective if they are not process. Moreover, since the contents of N.P. and K in three Ap's are, the grass paste could serve as manure in fish ponds also.

Three Ap's" - Livestock, Poultry-fish

"Three Ap's" are also palatable food for various animals in integrated fish farms. They do not need processing or just a simple processing, so the rate of utilization is very high. 900—1000 kg of three Ap's can rear one piglet to an adult with a body weight of 60–70 kg with a little amount of wheat and rice brans. The excreta of one pig can be converted into more than 40 kg of fish. In Helei Fish Farm, water hyacinth is fed to ducks at a rate of 150 g/duck/day with a little amount of wheat and rice brans. This can save 10,000 Yuan in feeding 22,000 ducks in half a year. The average excrement of each duck is about 52 kg and can be converted into 3 kg of fish. Thus, three Ap's become fish yield through two trophic levels (*i.e.* feeding levels).

DYKE-OOND SYSTEM

This is a special pattern of fish-cum-crop integration. On some farms, much of the broad dykes is devoted to economic crops:mulberry bushes, sugarcane, fruit trees, tea and rape, etc, or crops, vegetables and grasses. With different crops, this system can be called "mulberry plotfish pond"; "sugarcane dyke-fish pond"; "fruit tree dyke-fish pond" "rice paddy-fish pond", etc. Among these, mulberry plot-fish pond is more popular. Most of dyke-pond systems are distributed in Pearl River Delta and Taihu Lake basin. According to the historical records, mulberry cultivation and fish farming could be traced back to the 5th century B.C., but they did not form a well-linked dyke-pond system until the 16th century. The dyke-pond system in Taihu Lake basin is similar to that in Pearl River Delta, which is related to the common factors of the nature and the society. Take mulberry plot-fish pond system as an example:

Ecological structure of "mulberry plot-fish pond" system This system has a perfect ecological structure. It includes mulberry cultivation, sericulture, silk extraction, and fish farming with silkworm faece, pupae and waste water. Mulberry is the producer; silkworm is the first consumer; fish are the second ones land ecological system. In fish ponds, there are four ways of energy flow:

i. Silkworm faeces are directly taken in by fish and part of detritus can be filtered by filter-feeding fish.
ii. The inorganic nutrients in silkworm faces are utilized by phytoplankton & heterotrophic bacteria and the biomass of

phytoplankton and bacteria in turn are eaten by filter-feeding fish directly or indirectly.

iii. The leftover of input and fish faces are decomposed by hydro microbes, releasing inorganic nutrients; and then, the same process occurs as in (ii).

iv. At the same time, pond silt which is composed of all kinds of sediments returns to the pond dyke and the new material cycle begins.

ECONOMIC EFFICIENCY OF FISH-CUM-CROP INTEGRATION

To Increase Sources of Feeds and Fertilizers

This pattern of integration could provide a considerable amount of feeds for fish farming. The source of feeds is stable. They are much cheaper and of high quality. The cost could be reduced by one third.

To Save Energy

The energy consumption of transportation of purchasing feeds and fertilizers could be reduced.

To Rationally Utilize the Labour

Fish farming is a seasonal work. The input of labour varies greatly. The labour in slack season can be used for crop production. Part time worker can be fully utilized. This offers more jobs and increases the income

To Set up Reasonable Ecological System

Pond silt is used to plant fodder crops which in turn are used to feed fish. Large amount of fish could be produced and pond silt is accumulated again. It's a cycle which can fully utilize the sunlight, land, pond silt and fertile pond water and can improve the ecological condition of a pond. According to some data, grass planting on pond dyke can reduce the erosion of soils by 57%. Fish-cum-crop integration in line with the local conditions means to combine the Chinese traditional aquaculture with modern technology.

"Rotation" of Fish and Grass

Owing to the non-continuity of fish farming production, most of nursery ponds and grow-out ponds have been left fallow for a certain period of time, *e.g.* a yearling rearing pond has half a year of fallow from Nov. to June next year. During this period, fish ponds can be used to plant green fodder or green manure crops and then to culture fish. This could lead to the full utilization of fish ponds.

The Method of "Rotation" of Fish and Grass

The "rotation" of fish and grass can be divided into single pond rotation and multi-pond rotation.

Single Pond Rotation

That means the "rotation" of fish and grass in the same pond. This method is usually adopted after transferring fingerlings to other ponds.

Drain the pond, expose its bottom to the sun, trim the pond dyke, smooth the bottom, dig a ditch for drainage and then sow seeds of rye grass or barnyard grass or rice, etc. on the bottom and the slope.

The seeds of rye grass are broadcasted in Nov. at a rate of about 2 kg/mu while the seeds of barnyard and rice are generally sowed in late Apr. or early May in Jiangsu Province.

In order to extend the growing period germinated seeds could be sowed in the 1st 10 days of Apr. at a rate of 5–6 kg/mu. After sowing, get rid of birds and control the water depth and adopt different field management towards different plants.

Rye grass is a dry crop which can't endure waterlogging. Barnyard grass seedlings like to grow in shallow water with a depth of about 5–6 cm. During their growth period, they can be mowed as green fodder crops. Ryegrass can be mowed four times, with the yield of 3000–5000 kg/mu; barnyard grass and rice can be mowed once with the yield of about 2000 kg/mu. The last ratooning plant of the green fodder or green manure crops should be submerged for fermentation as manure.

But the amount of grass per square meter should be controlled within 4–5 kg. Then add water to about 2 meters high. In this way, it won't cause lack of oxygen in the pond water. 11–15 days later summerlings are stocked. After stocking, keep close watch over the water quality and fill the pond with fresh water or change the pond water in time accordingly. Barnyard and rice can be mowed and submerged by stages on the basis of their growth and then different summerlings can be stocked by groups.

Multi-pond Rotation

It needs a set of fish ponds, of which, a few are used to grow grass or green manure crops. Here are two ways. Because of short period of fallow of grow-out ponds, a few of fish ponds are drained first and planted with grass as early as possible but stocked with fish later than the other ponds, thus, extending the growing period of the grass; the rest are drained later but stocked with fish first. The fish ponds are rotated to do this each year. If there is a food constraint for fish, one or two fish ponds could be used to grow crops to provide feedstuff for fish in the other fish ponds. The fish ponds are rotated to grow grass each year.

THE MATERIAL LINK OF "MULBERRY PLOT-FISH POND"

The link of energy flow in "mulberry plot-fish pond" system is pond silt and feeds.

Pond Silt

Pond silt is the main source of manure for crops on the pond dyke. Each mu of fish ponds can provide pond silt for 1–2 mu of plants on dyke, usually for direct application but making compost in paddy-fish ponds in Taihu Lake basin. In winter pond mud (much mud, less water) is removed from fish ponds after draining and applied between mulberry bush lines on pond dyke or bailed (with a little water) to pond dykes and spread evenly after dry. After that, winter crops are intercropped between mulberry bush Lines.

In summer and autumn, liquid silt known as "nihua" (less mud, much water) is bailed to the pond dyke after mulberry leaves are picked 1–2 times, which is practised 2–3 times in Pearl River Delta and 5–6 times in Taihu Lake basin respectively every year to increase the soil fertility and when the base thickness of the soil of mulberry plants increase by 5–6 cm, it is beneficial to next crops. The mud on pond dykes are washed into fish ponds by rains; pond silt is supplied to pond dykes with nutrients. The process runs again and again.

Feeds and Fertilizers

Mulberry plots can supply feeds for fish directly or indirectly:

a. To provide feeds directly, including pasture grass, vegetables and mulberry leaves.

   On mulberry dykes, grass and vegetables can be planted after autumn leaves being picked; grasses and vegetables could be harvested at a rate of 3000 kg/mu and they can be converted into about 100 kg of herbivorous fish and 40 kg of other fish.

b. To provide feeds indirectly.

The first one is silkworm wastes which are a mixture of silkworm faeces, worm sloughs and mulberry leave residues. Silkworm wastes are rich in nutrients: organic material 87%; N 2.2–3.5%; P_2O_5 2.0–2.5%; K_20 1.5–2% and several trace elements but with less moisture, so the nutritive contents of silkworm wastes is higher than that in any kind of livestock or poultry manures. They serve as both fertilizers and feeds. One mu of mulberry plot can produce 2400–2500 kg of mulberry leaves, even 5000 kg. 100 kg of mulberry leaves when fed to silkworm can produce about 30–50 kg of silkworm wastes; therefore, 1250 kg of silkworm wastes can be obtained from sericulture with the mulberry leaves from one mu of mulberry plots. The conventional FCC of silkworm wastes in fish ponds of polyculture is 8. Hence, 1250 kg of silkworm wastes can be converted into over 150 kg of fish. The next one is pupae, which are by-products of silk extraction of cocoons. The protein content of pupae holds 55.8%, crude fat 29.1%. Pupae therefore are good feedstuff for feed-eaters.

Its FCC is 1.5–2. In Pearl River Delta, the production of cocoons from one mu of mulberry field is between 160 and 175 kg out of which about 130 kg of pupae can be obtained, and used as fish feeds, producing 90 kg of feed-eater spp. and their assorted spp.

The third one is waste water with large amount of protein detritus and soluable protein, which is obtained when the cocoons are steamed and processed at the factory. According to practices of the masses, 200 kg of waste water can be transformed into 1 kg of fish. More than 2500 kg of waste water can be obtained from the processing of cocoons which are gained from sericulture with mulberry leaves of one mu. Thus, 15 kg of fish can be produced. When the ratio of dyke and pond is 1:1, the total products (except pupae) from dyke can meet the demand of feedstuff for the fish pond with net fish yield of 300 kg.

INTEGRATED MANAGEMENT OF FISH-LIVESTOCK-POULTRY FARMING

The model of fish-cum-duck integration is rather common in China. Both land and water surface are the habitats of ducks. Fish-cum-duck integration is to utilize the biological relationship between fish and duck, which is of mutual benefit. It's not only beneficial to the fattening of ducks but also beneficial to the fish farming by providing more organic manures to fish. Thus, the fish yields can be increased. It's apparent that fish-cum-duck integration could result in good economic efficiency of fish farms.

The History and Status Quo of fish-cum-duck Integration

Duck raising in fish ponds is an ancient practice in Asia and Europe and this has long been practised in China too. The ancient farmers probably didn't know the advantageous effects of duck raising on fish culture. It's more likely that this integrated production pattern was formed incidentally on the basis of swimming propensity of ducks.

In 1934, the German scientist Probst conducted scientific experiments of integrated fish-cum-duck farming for the first time. He raised ducks in Common carp fish ponds and found that 0.9–1.7 kg of fish could be increased by raising one duck. Since the original output of fish in the experimental fish pond was rather low at that time, the increment of fish yields was conspicuous after ducks were raised in the pond; however, the outbreak of World War II kept the results unemployed. After World War II, the food was in shortage, especially animal protein.

The short supply of animal protein urged the development of the intensive commercial fish farming. Hungary, Czechoslovakia, GDR began to conduct integrated management experiments of fish-cum-duck in large scale early or late from 1952 to 1955. So far, integrated management pattern of fish-cum-duck has chiefly been practised in China, Hungary, GDR, Poland and Soviet Union. Fish-cum-duck integration has developed into a fixed model of integrated fish farming. In recent years, this model either on the scale aspect or on the managerial aspect has been developing very rapidly, especially in the area of a network of rivers like Jiangsu and Zhejing Provinces.

Beneficial Interactions in Fish-cum-duck Polyculture

From biological standpoint, a fish pond is a semi-closed biological unit. In fish ponds, there are many aquatic animals and plants, most of which could be used as natural food organisms of fish; some are predators of fry and fingerlings or detrimental to fish but can be utilized by ducks. If duck raising is conducted in fish ponds, the water surface can also be put in full utilization. Fish ponds provide ducks with an excellent environment, which prevents them from infection of any parasitic and other diseases. Ducks like to eat juvenile frog, tadpole, larvae of dragonfly.

That means to eradicate predators for fry and fingerlings. Furthermore, the protein content of these natural food organisms of ducks is high. Therefore, duck raising in fish ponds can appropriately reduce the demand for protein in duck feeds. For ducks raised in pens, the digestable protein content in duck feeds must be 16–18%, even higher up to 20% while for ducks raised in fish ponds, the digestable protein content might be reduced to 13–14%. This can save 200–300g available protein each duck, an equivalent to 2–3% of duck feeds. If the water surface is employed to raise ducks, the droppings of ducks directly go into water bodies providing C, N and P elements frequently. Thus, the biomass of natural food organisms in fish ponds can be increased. Here are two merits:

First, the droppings of ducks go directly into the pond water so that the availability of manure will not be lost at all; second, the "manuring" conducted by ducks on the surface is more homogeneous without any heaping of duck droppings. For this reason, duck raising in fish ponds can promote the growth of fish and increase the fish yields; in the meantime, it can naturally eliminate a series of problems which might be caused by pollution of duck excreta in duck pen.

Duck droppings are good organic manures in pond fish culture. The organic substance content in duck droppings are 3–5 times that of human excreta, inorganic substance such as N, 1.5–2.2 times; K_2O, 2.6–3.1 times. It shows that the quality of duck excreta is higher than that of human excreta. However, the quality and quantity of duck excreta are determined by duck species, feeds given, culturing management as well as climatical conditions, etc. In Europe, the stocking rate of ducks is generally 300–500 ducks/ha in summer and each duck produces about 7 kg of droppings during the fattening period of 36 days.

If 500 ducks are raised, the total duck excreta would reach 3000–3500 kg in that period. The moisture content in duck excreta holds 56.6%; organic substance, 26.2%; C,10%; P_2O_5, 1.4%;N,1%; K_2O, 0.62%; Ca, 1.8% and some other elements. The annual excreta of each egg-laying duck is 7.5–10 kg in dry weight, about 70 kg in wet weight. Each egg-laying duck raised in Helei fish farm, Wuxi, produces 40–45 kg of manure in wet weight per annum.

Besides the above mentioned, fish-cum-duck integration is beneficial to fish farming:

1. It can fully utilize the uneaten feeds because 10–20% of feeds are spilt by ducks to the ground. Based on the investigation, the fine feeds lost is 23–30g per duck per day or even higher. These spilt feeds and leftover falling directly into the pond water or flushed into fish ponds are utilized as fine feeds for fish resulting in achieving relevant fish yields. Furthermore, it reduces the wasting of duck feeds.
2. It can promote the recycling of nutrients in pond ecological system. The water surface is usually the duck's habitat. In shallow places, a duck often puts its head into the pond bottom and turns the silt to search for benthos. By virtue of this digging action, the nutritional elements deposited in the pond humus will diffuse. It's beneficial to the material recycling in the pond.
3. Ducks are the volunteer aerators. The swimming, playing and chasing of ducks in a pond more or less play a role in aeration, disturbing the smooth surface of the pond in reality.

Barash et al (1982) carried out an experiment on fish-cum-duck integration, rearing Common carp, Tilapia, Silver carp and Grass carp in an experimental pond of 400 m^2. Ducks were fed with nutrient-balanced feed pellet. At the beginning of the test, duck manure was applied at a rate of about 85 kg/ha in dry weight. At the end of the test, at a rate of 95 kg/ha in conformity with the growth of ducks. There was no other input of manures and feeds except duck excreta and spilt feeds. Compared with the duck raising in pens, the growth rate, feed efficiency and vitality and cleanness of eiderdowns and skin of the ducks raised in fish ponds are better than that in duck pens. Daily fish yield reached 38.5 kg/ha. .

Duck raising in fish ponds compared with duck raising in duck pens has three advantages:

1. The feed efficiency and the body weight of each duck increased. The higher feed efficiency also implies the spilt feeds uneaten by duck could be utilized by fish. Thus, the food conversion rate of fish-cum-duck integration was reduced from 3.84 to 2.64.
2. The survival rate of duck raising in fish ponds increased by 3.3% because fish ponds provided a clean environmental conditions for ducks.
3. Without any other feeds, only the droppings and leftover of ducks made the daily fish output of 36.5 kg/ha. However, someone thinks that if fish and ducks are raised together in the same pond, ducks will eat small fish. Based on many years practices in Helei fish farm and the introduction of duck raising experts, it's easy for ducks to

swallow a fish with a body weight below 4 g. Fish with a body weight above 5g have a strong ability to swim and can get away from ducks. In Helei Fish Farm, the recovery rate does not decline when Silver carp summerlings are stocked in fish-cum-duck ponds. Of course, the fry ponds and yearling ponds are not suitable to raise ducks because fish are too small and in high stocking density. Duck raising is generally practised in fingerling ponds or grow-out ponds.

FISH-CUM-PIG INTEGRATION

Fish culture combined with pig raising is a traditional integrated fish farming model in China. Now, a few of other countries are also engaged in fish-cum-pig integration. Since 1974, Dr. Buck has conducted experiments on utilization of pig manure in fish farming. In 1977, he polycultured Cyprinidae in the test pond with Silver carp as major species. The output was 4585 kg/ha. From economic view point, pig-raising can be divided into three situations: loss, balance and profit. If it combines with duck raising and fish farming, not only the economic efficiency could be increased but also the social efficiency and ecological efficiency could be raised. The pig food are often leftover and residues from the kitchen, aquatic plants and agricultural products and their wastes. The pig excreta is in turn used as organic manure in fish ponds. It might keep the environment clean. Pork is the indespensable subsidiary food. Pig manure is of high quality by virtue of nutritional completeness. That's why farmers usually combine fish farming with pig raising and it becomes a common pattern of integration in China's rural area.

Methods of Pig Manure Application

There are two types of pig sty in China. One is the simple pig shed constructed on the pond dyke or over the water surface; the other is centralized hog house. Both types have their own merits and demerits. The cost of the simple one is lower and moreover, it's easy to apply the manure to fish ponds. Therefore, this type is more suitable to household or small-scaled farms of fish-cum-pig integration.

The excreta of pig can automatically flow or be flushed into the pond. It can save much labour. If the area of a fish pond is less than 8 mu, a pig sty can be set up on the pond dyke and then, pig wastes after flowing into the pond, can diffuse to the whole pond by virtue of winds and waves. If more than 30 pigs are raised in the same spot, the method of flowing by gravity is not suitable because the more the pigs, the more the pig manure.

More often than not the place near the pigsty will be heaped up with manure and the water quality will be partially deteriorated. If pig manure sinks to the bottom too much or flows into the pond too much through the pipe, it will also cause fish surfacing. In order to fully utilize the effectiveness of manure, it's necessary to pay attention to the method of applying. The manure from the centralized hog house is easy to be concentrated to a storage pond or a sedimentation basin. The amount of application can be controlled by various means. Centralized hog house is fit to be built in large-scaled integrated fish farms.

The manure after dilution can be spread along the pond dyke by manual labour or by small boat in small fish ponds. If the fish pond is large, it's better to use boat and mechanic apparatus so as to spread the manure evenly. Here are two methods: To use stern-thruster boat (outboard motor boat) for fertilization Hang the iron cage loaded with manures alongside the boat. The space between grids is about 2.0—2.5 cm. The capacity of the cage differs with the amount of fertilizers needed. The cage loaded with manure must be hanged 10—20 cm below the surface. When the engine starts, the current formed by the thruster will rush against the manure in the cage and as a result, it can be spread evenly.

To Install a Pump in a Boat

The manure should be diluted in a cabin through a funnel and then be spread into the whole pond.

The large quantity of liquid manure in sedimentation basin of largescaled integrated fish farm can be introduced through a pipe or a hose to the pond side and then be sprayed through a nozzle. The average rate is 200 litres/min each nozzle. The nozzle should be fixed 0.5—1.0 m above the surface, in this way, the liquid manure can be fully contacted with air. The nozzle will serve as an aerator if it works to spray water when the D.O.C. in pond water is low.

In fish-cum-pig integration, two points must be especially noticed: one is lack of oxygen in the pond; the other is oligotrophic pond water. That's why the water quality should be monitored at any time. Besides, the production period of pig should match the demand of pig manure in fish farming.

Integrated fish farms are chiefly engaged in raising hogs. The advantage is that the production periods of pig and fish farming match well. Hog can be raised for 2 periods a year. The duration of each period lasts for 5 to 6 months. 60% of the total amount of manures applied to fish ponds are given during the first half of a year. The peak of application is in Jan. and Feb. for base manure and in June and July for the additive manure. Usually, no manure is applied after the middle 10 days of Oct. Therefore, one pig production period is supposed to be from the middle 10 days of Feb. to the middle 10 days of Aug; the other is from the middle 10 days of July to the middle 10 days of Jan. The growth of the first batch of pigs peaks at the latter part of the production period and their excreta as well. It can just meet the fertilizer demand of fish farming. The excreta of pigs from Nov. to Jan. serve as base manure for fish farming next year. This collocation not only can meet the demand of pig manure during the whole period of fish farming but also satisfy the concentrated demand of pig manure in fish farming.

FISH-CUM-DUCK INTEGRATION OF FISH FARMING

In fish-cum-duck integration, the result of fish farming has much to do with how to raise ducks. There are three types of farming practices:

RAISING LARGE GROUPS OF DUCKS IN OPEN WATER

This is "grazing" type of duck raising. Average number of a group in grazing is about 1000 ducks by one worker. The ducks are generally let loose to go grazing in rivers, lakes and reservoirs during the day, utilizing natural food organisms, but are kept in pens at night. This method is advantageous to large water bodies for promoting fish proliferation and can be considered as an integrated management model for large water bodies, such as lakes and reservoirs, etc. As for those fish pond farms, this method of integration can not effectively utilize the duck manures except the increment of economic benefit from duck raising.

Raising Ducks on Pond Shore

A relatively large duck shed including a workshop for the administration is constructed in the vicinity of fish ponds with cemented area of dry and wet runs outside. The average stocking rate is about 4 ducks/m^2. The dry and wet runs are being cleaned once a day. During cleaning, the sluice of the wet run is opened to allow organic manure to be flushed into fish ponds through a manure ditch. After this, the sluice is closed and the wet run is filled with fresh water. This method has the advantages for centralized management and adopting mechanisms, but is unable to fully utilize leftover and undigested feedstuff in duck excreta. The effects of leftover and the direct value of feeds more or less are lost. It's also unable to take advantage of the symbiosis of duck and fish.

Raising Ducks in Ffish Ponds

This is the common method of integrated fish-cum-duck farming. The dikes of grow-out or 2-year-old fingerling ponds are partly fenced to form a dry run and part of the water area or a corner of the pond is fenced with used material to form a wet run. The net pen is installed by 40–50 cm above and below the water surface respectively in order to save the net material. In this way, fish can enter the wet run for food while ducks can not escape under the net. In a large pond, a small "island" is constructed at the centre of the pond for installation of demand feeding facilities. The stocking density currently practised in China is higher than that practised in other countries, averaging 4.5 individual/m^2 of pen shed including the dry run and 3—4 ind/m^2 for the wet run. In the early phase of integrated management of fish-cum-duck, ducks went everywhere in fish ponds to get food; now this pattern has already been improved. The duck raising area has been set up to connect the duck shed, the dry run and the wet run. Whether fish-cum-duck integration succeeds or not primarily depends on technical measures of duck raising. Both meat ducks and egg laying ducks can be raised in fish ponds. In summer, the 14-day ducklings are accustomed to the life on the water surface. Meat ducks grow fast. The good stock can reach to the marketable size of 2 kg in fish ponds within 48—52 days; the stocks which grow slowly can reach to the marketable size within 55—56 days. Ducks should be on the market as soon as they reach the marketable size or they will lose their feathers, resulting in decrease of food efficiency, body weight and commodity value.

The number of ducks to be raised in this type of polyculture depends on the quantity of duck excreta which in turn are determined by the duck species, quality of feeds given as well as the method of raising them. In raising Beijing ducks, about 7 kg of duck excreta per duck can be obtained during 36-days fattening period. The egg laying Shaoxin ducks raised in Wuxi produce 42.5—47.5 kg of manure per duck per annum, while the hybrids of Shaoxin and Khaki-Combell ducks produce more than 50 kg per duck.

The stocking rate of ducks also depends on the climatical conditions and the stocking ratio and density of the various fish species polycultured in the pond. In Europe, the stocking rate is generally 500ind/ha. As a result, the increment of fish yield will be 90 kg/ha. In tropic and sub-tropic zone, Woynarovich (1980) recommended that the stocking rate should be 2250 ind/ha. In Hongkong, the optimum stocking rate is 2505—3450 ind/ha. In Wuxi, 2000 ind/ha. If the meat ducks are raised in fish ponds, the number should be reduced in view of larger quantity of excreta. In Taihu district, polyculture of 7—8 species of fish is practised in fish ponds, in which the stocking ratio of various species of fish remains unchanged when ducks are raised. If the number of ducks exceeds 3000 ind/ha, then the filter-feeding fish and omnivorous fish could be increased and herbivorous fish could be reduced.

The stacking of organic materials won't take place in this type of integration so long as the stocking rate of ducks is proper and the excreta amount does not exceed the transforming power of a fish pond. Ducks swim loosely in the wet run to search for food, their excreta drop evenly into the wet run and the effects of fertilization of duck droppings expand to the whole pond through winds and waves.

Economic Efficiency

The integrated management of fish-cum-duck farming especially raising ducks on the surface of fish ponds is an economically efficient farming practice. In 1980, Helei Fish Farm conducted a comparative test between pond no. 13 and pond no. 21. Two ponds were adjacent with the same size about one ha. There were pigsties set up on the dyke between the two ponds. Pond no. 13 was stocked with 2207.5 kg of fingerlings and 1900 egg laying ducks without any input of manure; Pond no. 21 was stocked with 2188.5 kg of fingerlings only. The species and size of fish and the feeds given in Pond No. 21 were the same as those in Pond No. 13 but with input of manure at a rate of 20,000 kg/ha/year. The output of Pond No. 13 was 12,234 kg while Pond No. 21 was 10,464 kg. From the above mentioned comparative test, if the stocking rate of egg laying ducks is 1830—1920 ducks/ha, apart from economizing input of 300000 kg of manure, per ha fish yields can be increased by about 17% over ponds without duck integration. Based on the data and calculation, the average rearing period of ducks is 10 months in a year and about 2.5-5.5 kg of fish might be produced by raising one duck. According to the test in Helei Fish Farm, about 2.5 kg of fish could be produced besides about 200 eggs, (260—300 eggs for the hybrids) by raising one duck in fish pond. Helei Fish Farm raise 20000 ducks every year, providing 850000 kg of duck manure. In 1980, the farm got a net profit of 42000 yuan from duck raising. There were 48 farmers in the farm. The average income was 881.25 yuan per capita.

It's impossible to get an accurate economic analysis on fish-cum-duck integration because the investment, production cost and the yields of duck and fish varies in different countries; even in the same district, fish and duck species, stocking densities, quality and efficiency of feeds, rearing management and climatical conditions differ in thousand ways.

It can be explained by an example:

In 1981, Helei Fish Farm raised 22000 ducks. Apart from providing 1,000,000 kg of duck manure and large amount of leftover to fish ponds, it harvested 212,695.9 kg of duck eggs, 6059 kg of duck meat, thus providing the market with 24315 kg of available animal protein which is equivalent to 215560 kg of Grass carp protein. The total annual income was 57740 yuan including 10800 yuan from ducks, 42000 yuan from duck eggs, 4940 yuan from duck excreta for fish farming and it holds 24.62% of total profit of the farm.

From the view point of input-output relationship, integrated fish farming with ducks is considered to be the best model in integration of fish, livestocks and poultry. The economic efficiency of fish-cum-pig integration from micro-economic stand point is generally not so high and its profit is low. In the case of integrated fish farming with poultry, fish-cum-chicken integration lacks symbiotic relationship, which exists in the case of integrated fish-cum-geese farming, but the egg-laying rate of geese and the market demand are rather low. By comparing the protein input and output in integrated fish-cum-duck farming, it's found that the production of 1g of egg protein from Shaoxin ducks requires 5.53 g of feed protein whilst 1g of dairy protein from cows needs 5.55g of feed protein. They are similar in weight; however, it is relatively easier to raise ducks than cow and the economic efficiency and income of fish-cum-duck far exceeds that of fish-cum-cow, *e.g.* in 1981, a worker in Helei Fish Farm produced 292.6 kg of protein from cows while his counterpart produced 506.6 kg of protein from ducks which is 73% higher. The net profit per capita in fish-cum-cow farming is 1067.9 yuan and 1427.08 yuan in fish-cum-duck farming, which is 33.6% higher.

Integrated management of fish-cum-duck farming can be further developed to achieve higher economic efficiency by utilizing the natural water body to cultivate high-yield aquatic plants as vegetable feeds of ducks and by utilizing the wastes of integrated fish farming and from city proper to grow earthworm as animal-based feeds for ducks. At the output end, the products such as fish, ducks and eggs could be further processed before marketing, thus, the economic efficiency and income can be considerably raised either from the angle of energy and nutrient source or economy.

RATE OF PIG RAISED PER UNIT AREA OF FISH PONDS

The actual rate of pig raised for per unit area of fish pond in integrated fish farms in different areas is from one to five per mu of fish pond. How much excreta of pigs can the given unit area of fish pond accept so as not to deteriorate the water quality of the pond but to maintain sufficient food for fish? It must be worked out through the productive experiments in line with the local conditions. In China, a fish pond is often given various kinds of organic manures such as pig manure, cow dung or other animal manures so the quantity of applying pig manure must be reduced correspondingly.

The capacity of a fish pond to accept animal manures depends upon the techniques of application and the nature of manures. In Hungary, the growth period of fish is 150—180 days. If pig manure is applied by the method of stacking in a corner, the rate of application is 100—134 kg/mu of fish pond. If the method of so-called carbon manuring technique is adopted, that is to say, fresh pig manure mixed with pond water is spread to the whole pond frequently, the rate of application is 20—40 kg/mu/day. It's an equivalent to 1000–1500 kg of condensed liquid manure or 1.2—2.5m^3 of solid wastes of commercial pigsty.

In theory, the maximum capacity per mu to accept manures is 2 to 3 times the above mentioned figures. Why, it involves many facets such as environmental conditions of a fish pond, the quality of pig manure, the managerial techniques of fish ponds and so on.

Fish Species Cultured in Fish-cum-pig Integration

The production efficiency of fish-cum-pig integration appears in the full utilization of food organisms by fish in pig manured pond. In 1950's, when Hungarian began to culture fish by using pig manure, two problems happened. (1) Water bloom of Aphanizomenon flosaquae and other blue algae appeared time and again because of non-control of manure application. It causes the disappearance of Daphnia spp.

Common carp can not utilize plankton directly, resulting in waste of primary productivity, The filtering species should be stocked in manured pond, and thus, it will avoid auto shading of phytoplankton and decreasing the products of photosysnthesis. For this reason, Hungarian have stocked certain kinds of Chinese carps since 1970. Silver carp and Bighead are the best species to control plankton. They have fine filtering organs and can filter and eat phytoplankton in size above 220ì. If the water temperature goes up to 22°C, they can filter and eat more plankton and a great number of detritus and bacterial congregations in a given unit time. The feed basis formed after fertilization could be utilized more. In mono-pig-manured pond, a few Grass carp may be stocked to control Aphanizomenon flosaqae and submerged vegetation in the pond. In mono-culture of Common carp, 50 kg of pig manure can be converted into 1.25–1.5 kg of Common carp; in polyculture with Common carp as major species, 50 kg of pig manure can be converted into 1.75–2 kg of fish; in polyculture with Silver carp as major species, 50 kg of pig manure can be converted into 3 kg of fish. The feeding habit of Tilapia is wide in scope and serves as cleaner in pig manured pond. This species fits the pigmanured pond. In Hubei Province, China, Silver carp, Bighead and Tilapia are the main species in pig manured polyculture, assorted with a few Grass carp, Common carp and Xeno cyprinus as minor species.

Economic Efficiency of Fish-cum-pig Integration

The economic efficiency is apparent if the wastes of pig raising can be utilized. In Hunan Province, one mu of fish pond matches with three pigs. In polyculture with Silver carp and Bighead as major species, the output could reach 150–200 kg/mu. In tropic and sub-tropic zone, the temperature is higher and the biological process becomes quicker, more pig manure can be decomposed in the pond and moreover, with proper species in polyculture, the utilization rate of pig manure will perhaps be much higher.

Fish-cum-pig integration can reduce the production cost. For example, in 1981. Xinan Fish Farm Wuxi gained 3388000 kg of pig manure including 50% flush water. It saved 14229.6 yuan of manure expenditure. The fish yield was 33880 kg. The profit of pig raising itself was not too much, nevertheless, the total revenues of integrated fish farm increased because the excreta of pig took the place of artificial feeds and inorganic fertilizers, which accounted for 58.8% of the cost of fish farming.

However, the simple fish-cum-pig integration becomes less in China. This pattern is usually a part of fish-livestock-agriculture integration system which is a comlete man-made ecosystem.

Fish-Cum-Cow Integration

Fish farming using cow manure has long been practiced in China. Since the founding of the People's Republic of China, cow farming has developed rapidly. It promotes fish-cum-cow integration, which is one of the common modellings of integration in the southern part of China. On fish farms cow farming can save much fertilizers, cut down fish feeds and increase the income from milk while fish farming on cow farms can make it realized that "the excreta of a cow can be converted into a thousand jin of fish". Moreover, cow manure can be disposed at hand resulting in saving much money, labour and energy and also it is good for the environment. If farmers use part time worker to conduct small scale integrated fish farming with cow raising, they can not only earn more money but also can supply both fish and milk to the market.

MANAGEMENT OF FISH-CUM-COW INTEGRATION

In the integrated fish farm, cow sheds should be built in the vicinity of fish ponds so as to utilize cow manure at hand. The faeces and urine are collected separately, the former transported by convyor or boat or car and spread evenly, the latter pumped into fish pond.

If the floor of cow shed is higher than the pond dike, manuring ditch could be dug to collect faeces and urine together, then to flush them into fish ponds. It saves time and labour. But the above methods often cause stacking of manure and uneven fertilization. The best way is to mix up cow dung and urine and to pump them into fish ponds evenly. If the hose pipe has a spray nozzle, the result will be better. If the area of fish ponds is large, the boat dragging a stern manure bucket could be used. Then manure will be evenly spread by the waves caused by sailing.

Proportion here means the area of fish pond matched with one milk cow. It depends on synergism of many factors, such as the amount of cow manure and wasted food; the species ratio and target output of fish; the sediment of pond; the quality and quantity of pond humus; other manures and feeds applied. Supposing fish culture wholly depends upon cow manure and wasted food of cows with a net production target of 250 kg, among which 10% is herbivorous fish, the rest filtering and omnivorous fish, each milk cow can provide manure and feeds for fish in 2 mu of fish pond. If the proportion of feed-eater increases, the method of calculation of fish-cow proportion is shown in the empirical equation related to manure demand and quantity of livestock and poultry in model plan.

Period and frequency of manuring stands on the transforming speed of cow manure and season change and the rule of the ebb and flow of natural food organisms in fish pond. In 1983, Freshwater Fisheries Research Centre carried out an experiment on measuring the peak time of N and natural food organisms at different water temperature in a fish pond manured once a week. It showed all nutritive factors had two peaks in a week. The high peak appeared earlier when the water temperature rose. Therefore, the guidelines to determine period and frequency of manuring are as follows: In late winter and early spring, with sufficient base manure, a few times but in large quantities, that is once every 5–7 days; in spring and autumn, once every three days; in summer, once every day or every two days in small quantities and it's better to manure everyday and make some adjustment according to weather, water colour and fish growth.

Stocking pattern in fish farming by applying the waste in cow raising Cow dung and urine is beneficial to filtering and omnivorous fish culture. Therefore, Silver carp and Bighead are taken as major species assorted with omnivorous fish (Common carp as main minor species), and 15—20 % of herbivorous fish. If it's over that ratio, supplemented feeds must be applied. It's optimal that the output of herbivores occupies around 12% of total net output of the pond.

THE MATERIAL BASIS OF FISH-CUM-COW INTEGRATION

In fish-cum-cow integration, cows are suppliers of the materials for fish farming. They can provide cow manure, leftover and matted grass of cow sheds, etc.

The Biological Basis of Fish Farming using Cow Manure

Among all the excreta of livestock, the amount of excreta of cows is the greatest and the most stable. If a cow is 460 kg in body weight, it can produce 0.081 kg of manure/kg of body weight per day and the annual amount of faeces can reach 13600 kg and that of urine about 9000 kg.

The nutritive content of cow dung is a little less than that of pig faeces. . If 0.024 kg of fresh cow manure is applied to a water body of 1 cubic meter per day, inorganic N and P in fish ponds will be 0.897 mg/L and 0.024 mg/L respectively, which are close to the inorganic N, 0.97—2.06 mg/L (average 1.38±0.42 mg/L) and the inorganic P, 0.018—0.036 mg/L in high-yield fish ponds. The ratio of N/P will be 36.9, which is a little higher than the value of the ratio of N/P in phytoplankton. Apparently it will limit the large increment of phytoplankton in fish ponds. Nevertheless, the average amount of phytoplankton in manured pond can still reach 19.15±6.5 mg/L which is apparently higher than that in control ponds and close to the lower limit of optimum food density (20—100 mg/L) of Silver carp and Bighead. The average biomass of zooplankton will amount to 5.61 mg/L which is much higher than that in unmanured pond and the ratio between the biomass of zooplankton and phytoplankton is 1/3.4.

Both the biomass of zooplankton and the ratio between the biomass of zooplankton and that of phytoplankton come to the level as much as in fertile water fish pond (that is to say, the biomass of zooplankton is between 5 and 30 mg/L and the ratio is between 1/4 and 1/3).

No Risk in Fish Farming using Cow Manure

Cows are ruminants. It is due to repeated grinding and digestive decomposition catalyzed by great amount of micro-organisms in rumina that cow manure is very fine. It can suspend longer in the water. The sinking speed of cow manure particles is 2.6 cm/min and that of pig manure particles 4.3 cm/min. If the same amount is applied, the sediment of pig manure is 33%

higher than that of cow manure after 24 hours while the suspended organic detritus below 0.65u in size in cow-manured pond is 39.99 mg/L, which is 153.4% higher than that in pig-manured pond. And it occupies 54.6% in total amount of suspended particles, which is the highest percentage compared with that in pig, duck, or chicken-manured ponds.

The suspensibility not only enables fish to get more feeds but also reduces oxygen consumption caused by manure stacking and avoids forming of harmful gases. The BOD of cow manure is lower than that of other livestock manures because the cow forage has already been decomposed by micro-organisms in cow's body. The BOD of 1 kg of cow manure in five days is 20.6g, 32% lower than that of pig manure which is 30.0g. The same tendency appears when they are in fish ponds.

In one week's consecutive measurement, the lowest dissolved oxygen content in cow-manured pond at 5:30 a.m. is 1.8 mg/L while that in pig-manured pond is 1.0 mg/L. The average dissolved oxygen content is 4 mg/L in cowmanured pond but only 3.3 mg/L in pig-manured pond. By virtue of suspensibility and low oxygen consumption of cow manure, fish farming using cow manure is more safe. Freshwater Fisheries Research Centre conducted an experiment on different livestock manures in fish farming in 1983. It indicated that the survival rate of fish in cow-manured pond is 98%, second to none.

Wasted Food of Cows for Fish Farming

Cows feed chiefly on grass. During the grass growing season of about 7 months, the fresh fodder grass for each adult cow amount to 9000—11000 kg. But the leftover occupies 27.9%, around 3000 kg.

However, that period of time happens to be the highest ingestion seasons of herbivorous fish, therefore, the wasted fodder of cow can be utilized as fish feeds. The results differ with different quality of grass. The food conversion coefficiency of terrestrial wild grass is about 40–50. The matted grass in cow shed can only be used as compost of the pond but the leftover of fine fodder for cows can be used as fish feeds.

ECONOMIC BENEFITS OF FISH-CUM-COW INTEGRATION

Both Milk and Fish Harvests

Each black-and-white milk cow after three births can produce around 5000 kg of milk annually, which contains 155 kg of protein. Except the milk for calves, each cow can not only supply the market with 4500 kg of milk, but also can provide feeds and manures for fish farming, which in turn can produce 500 kg of fish containing 55.8 kg of protein annually. Each cow can provide a total of 210.8 kg of protein that is 3700,000 kcal in calorie value annually. Fish-cum-cow integration can increase protein by 36 % and calorie value by 10.6 % above the unitary cow farming and it can increase protein by 2.7 times and calorie value by 9.5 times over those produced by unitary fish culture.

Increasing Revenues, Decreasing Expenditure

Fish-cum-cow integration can get more money from fish, milk and calves. The output value of each milk cow is averagely more than 3100 yuan. The excrements and leftover of one cow can be converted into 500 kg of fish without additional feeds and manure. Therefore, it can reduce half of the cost of fish farming.

Alleviating Unemployment

In fish farming, raising four cows can offer one opportunity of employment for people.

Saving Energy

Fish-cum-cow integration can provide feeds and manure for fish farming so that it can save a lot of energy for transportation. However, the investment of fish-cum-cow integration is much greater at one stroke. It needs about 2500 yuan RMB for each milk cow and attached facilities. On the basis of investigation, the profit rate of cow is higher, which is about 30%, that is to say, 100 yuan net profit can be gained from 300 yuan fixed capital. In term of net profits and cost saving in fish farming, the economic return of investment will cover less than three years.

ECOLOGICAL EFFICIENCY OF PIG-GRASS-FISH INTEGRATION

To Raise Utilization Efficiency of Light energy

The products of photosynthesis depends on the amount of light that is the product of light intensity by sunshine time. Terrestrial pasture grasses can absorb light energy from the sun directly and can avoid shading each other under artificial control, *e.g.* mowing in time, so as to reduce the reduction rate of radiant energy.

The rate of reduction of radiant energy in water is higher than that on land because the water surface reflects and scatters light. When the sun rises high, the light intensity will be reduced 6% because of reflection of calm surface or 10% because of reflection of waving surface. When the sun rises or sets, most of light can not enter into water due to the shading of pond sykes, therefore, so far as the sun light is concerned, daylight in water is shorter than daylight on the land.

If the gradient of a pond dyke is 30 degree the sunshine time of the pond water along the pond dyke is reduced by one third in comparison with that on top of pond dykes.

The lower the water level and the smaller the pond area, the shorter the sunshine time. Since there is a large amount of suspended and dissolved substances in fish ponds and especially plankton propagate greatly, auto-shading occurs and it's hard to control. According to the experiments of

Freshwater Fisheries Research Centre, the transparency and the amount of suspended substances are in inverse proportion. In a natural pond, the coefficient of correlativity r=-0.9083. The regression equation is as follows:

$$Y = 247.692 - 2.840\ x$$

Y = transparency (cm)

X = the amount of suspended substances (mg/L)

Larcher (1980) realized that radiation intensity is in inverse proportion to the water depth, decreasing in the form of exponent (or index). When the sky is evenly clouded, the light intensity, according to synchronous measurements, on the soil surface with the depth of ryegrass of 40 cm, is 900 lx, which is 13.43% of that on the surface of grasses and that means its relative light intensity reaches 1.8% of 50 klx, the saturated point of net photosynthesis of land grasses; the light intensity at 40 cm below the water surface of a pond is 28 lx, the relative light intensity is only 0.42% of that on water surface. It only reaches 0.14% of 20 klx, the saturated point of net photosynthesis of phytoplankton. Moreover, 40 cm is the maximum height of ryegrasses to be mowed. In the growing or ratooning phase before mowing, the relative light intensity is much higher than the measured one while the relative light intensity in the pond cannot be increased artificially. Therefore, from the view point of light intensity and sunshine time net photosynthesis of phytoplankton cannot compare with that of terrestrial plants either in a specific time or in a period of time.

Terrestrial grasses have greater ability to photosynthesize than phytoplankton. Under the normal supply of carbon dioxide and with an optimal light, temperature and water conditions, the uptake of carbon dioxide of sunplant herb is 20–50 $mg/dm^2/hr$. The intake of energy and the primary productivity of terrestrial grasses are higher than those of phytoplankton. Manured with pond silt alone, the yield of ryegrass can reach 4500 kg/mu in Apr. and May. The average daily output is 82 kg in May. The average daily yield of Sudan grass can reach more than 100 kg/mu.

The net output of oxygen produced by phytoplankton in Wuxi or Xiamen is 8–10g $o_2/m^2/day$. So the net yield of phytoplankton is only 42.3–52.9/ma/ day. (Le xing-zhi et al). From the measurement in Wuxi high yielding fish pond, the net yield of phytoplankton in peak period was 41 kg in wet weight per mu per day, only 50% of the net yield of rye grass, 41% of the net yield of Sudan grass per mu per day.

After calculation, in accordance with the balanced equation of photosynthesis, 1 kg of O_2 is equal to 6.1 kg of phytoplankton in wet weight while 1 kg of O_2 means producing 3.51 Mcal of energy. Therefore, 1 kg of phytoplankton averagely holds 0.575 Mcal of energy. Daily output of plankton from May to Aug. is 41 kg/mu, that is, 23.575 Mcal. 1 kg of ryegrass holds 1.035 Mcal. Daily output of ryegrass in May holds 84.87 Mcal, that is 3.6 times that of phytoplankton. 1 kg of Sudan grass holds 0.75 Mcal. Daily output from June to Sept. is 77.25 Mcal, which is 3.3 times that of phytoplankton.

In the case of rye grass and Sudan grass. We can see that the primary productivity of terrectrial grasses is 2 to 4 times that of phytoplankton in high yielding fish ponds and the output of energy 3.3–3.6 times. The yield of terrestrial grasses can be further increased through cultivation and popularization of good varieties of seeds, rotation systems and intercropping with leguminous grasses and scientific management. On the contrary, how to increase the primary productivity of phytoplankton is a hard nut to crack so far.

The utilization rates of primary productivity of phytoplankton and terrestrial grasses by fish are different. Furthermore, phytoplankton can continue their propagation only after part of seeds remain in a pond. He Zhi-hui (1983) calculated the consumption of phytoplankton in high yielding fish ponds in Wuxi. The direct phytoplankton consumption of Silver carp, Bighead, Tilapia and Crucian carp or through zoopalnkton accounted for 82% of the biomass of phytoplankton. (70% of the zooplankton are consumed by plankton-eater.); whereas the utilization rate of ryegrass with the height of 50 cm or Sudan grass below 1.2 meter by fish is 100%.

To Utilize Two Primary Productivity

Compared with manuring the pond water, pig-grass- fish integration utilizes high effective primary productivity of terrestrial grasses and at the same time, it also uses the primary productivity in fish ponds. In this system, herbivorous fish are the major species, which are stocked according to the stocking pattern with net yield of 400 kg/mu. The mean capacity load of herbivorous spp. is 190 kg. The daily Sudan grass ingestion of herbivorous fish occupies 22.5–30% of the body weight of fish, average 25%.

Thus, the ingestion of herbivorous fish is 48 kg/mu/day. The excreta of herbivorous fish are about half of its ingestion amount, so the daily amount of excreta are 24 kg. On the basis of analysis of South-Western Normal College in China (1977), the N and P contents of Grass carp excreta are 1.102% and 0.426% which are three times and two times those of pig excreta respectively. At least, the daily excreta of herbivorous fish are equal to 24 kg of net pig manure. During the 250 days of ingestion period, the amount of excreta of herbivorous fish are equal to 6000 kg of net pig manure or the amount of net excreta of 6 pigs. It's near to the total amount of manure application in the fish pond with Silver carp and Bighead as major spp.

Thus, growing grass by using pig manure can provide sufficient fodder grasses to the herbivorous fish, the excreta of which, in turn, can supply nutritive salts enough for the fish pond so that the primary productivity of the fish pond can reach the mean level of the pig-manured pond. In Wuxian, Zhangzhuan Aquaculture Brigade which has adopted pig-grass-fish integration, all the pig manure are used to grow grasses. No additional manure is applied to the fish pond with herbivorous fish as major spp., which occupy

55–72% of the total stocking amount. The water colour looks green and brownish green with transparency of 10–35 cm. For example, in pond No. 33 with herbivorous fish occupying 72%, the average transparency is 17.5 cm. According to the regression equation of phytoplankton and transparency after He zhi-hui.

$Y = 153.69 - 3.06\ x$

Y = the biomass of phytoplankton (mg/l)

X = transparency (cm)

The biomass, of phytoplankton of that pond is about 100 mg/l. The biomass of phytoplankton of high yielding fish ponds in Wuxi is between 20 and 100 mg/l. It indicates that the excreta of herbivorous fish can fertilize the pond water and provide sufficient biomass of plankton demanded by plankton-eater.

Pig-grass-fish integration can avoid certain ecological defects caused by direct application of pig manure in ponds. The direct application increases the BOD and lowers the dissolved oxygen levels esp. if stacking application of manure is adopted at ten-day intervals. The BOD_5 = 29.09 – 30.75 g/kg. According to the experiments of Fresh water Fisheries Research Centre, the average BOD_1 of fish ponds are 4.10 mg/l and 5.48/1 respectively when pig manure is spread evenly into the pond at the rate of 0.12 kg and 0.29 kg/m^3/ week. At 5:30 in the morning, the average D.O. is only 1.0 mg/l. Zhangzhuan Aquaculture Brigade adopted pig-grass-fish integration in 10 ponds, and supplement molluscs and fine feeds without any other manures in fish ponds. The carrying capacity of 9 ponds were from 779.3 to 1350.5 kg/mu except one pond with 475.3 kg/mu. All Dissolved oxygen levels were high. The minimum DOL was 2 mg/l and average 4.6 mg/l in Pond No. 10 with the highest carrying capacity.

Because this system utilizes two primary productivity, avoiding some ecological defects, there is a greater potentiality of increasing fish yields. In terms of many rcports from abroad, the fish yields of filterfeeding species and omnivorous species in a manured pond only reach 15–31 kg/ha/day when they are monocultured or polycultured as major species. Generally speaking, the yield of plankton-eater is 300–400 kg/mu. It's difficult to surpass this figures and moreover, the economic efficiency will be reduced when the yield increases. Nowadays, in most of fish ponds with the yield of 500 kg, the herbivorous fish are cultured as major species with grass as main food and without application of manure or with just a little manure.

The excreta of 1 kg of herbivorous fish can fertilize the pond water and produce 0.2–0.5 kg of Silver carp and Bighead. The larger the size of fingerlings the more their ingestion, and the more the number of planktoneater that they bring along. In Grass carp rearing pond with gross yield of 110–150 kg compared with Grass carp rearing pond with gross yield of 50–100 kg, the net yield of plankton eater is 37.5 kg higher.

In fish-pig link, filter-feeding fish and omnivorous fish feed on detritus and the bacteria on detritus. The relationship between the fish yield and both detritus and bacteria depend on their quality rather than on their quantity, therefore, it can not be overestimated just according to the amount of detritus in a water body. The experiment of Lin Wan-lian also proves on one aspect that it is a loss to use pig manure to grow grasses for fish farming so far as part of detritus, which can be utilized by fish, is concerned. Nevertheless, the good effects of pig-grass-fish integration can make up for the loss. What's more, the effects of detritus is not so good.

MODELS OF FISH-LIVESTOCK-CROPS INTEGRATION

The model of fish-livestock-crop integration is the transverse combination of fish-cum-livestock and fish-cum-crop models, namely, to utilize animal manures and pond silt from fish-cum-livestock integration as agriculture manures and to utilize crop products from fish-cum-crop integration as feedstuff for fish and livestock. So far the most popular model is pig-grass (vegetables)-fish and vegetables-pig-fish integration, etc. The former develops more rapidly nowadays.

In fish-cum-pig integration, all pig excreta are used to fertilize fish ponds with plankton-eater as major species whereas in pig-grassfish integration, part or the whole of pig excreta are used as manure for terrestrial and aquatic fodder plants with high-yielding pasture grasses as major crops. Green fodder crops are used to culture herbivorous fish and bring along rearing filter-feeding fish. No or just a little pig excreta go down directly to fish ponds. Finally, pond silt accumulated in fish farming returns to the land as crop fertilizer. The new nutrient cycles and energy flow is thus set up in pig-grass-fish integration. Both terrestrial and aquatic primary productivity is fully utilized.

Therefore, its structure is more rational and more effective in comparison with fish-cum-pig integration. It can make up for insufficient manures in fish-cum-grass integration.

THE ECONOMIC EFFICIENCY OF PIG-GRASS-FISH INTEGRATION

On the basis of ecological principles and its good economic efficiency, the pig-grass-fish integration can get better economic efficiency than fish-cum-pig integration. In 1979, Zhangzhuan Aquaculture Brigade performed a productive test. 50 kg of pig exereta (pigsty effluent occupies 50%) or 50 kg of nightsoil can produce 56 kg of terrestrial grasses.

50 kg of net pig manure can produce 112 kg of grasses. The Food Conversion Coefficient of English rye grass and Sudan grass is 18.4–27.5, average 27. Thus, 100 kg of pig excreta through growing grass can be converted into 4.2 kg of herbivorous fish, bring along rearing 1.5 kg of filter-feeding and omnivorous fish, the total yield is 5.7 kg.

According to the tests of Freshwater Fisheries Research Centre, 100 kg of pig excreta can be converted into 2.5 kg of fish, whereas 50 kg of net pig excreta through growing grass can produce 3.2 kg of fish, increasing 133 %. In pig-grass-fish integration, the dominant species are herbivorous fish, which can meet the demand of the market and gain much profit. 100 kg of pig excreta are used to grow grass then to culture fish, resulting in the increment of 2.90 Yuan which is 1.34 times that of fish farming by applying pig excreta directly.

ALLOCATION OF PIG-GRASS-FISH

In the combination of pig-grass link, the guide line is to facilitate grass planting and manure application.

If grasses are planted on pond dykes only, the simple pig sty can be built near the pond; if there is a fodder field beside pond dyke, centralized pig shed near fodder field is taken into consideration.

Firstly, the ratio of pig and grass depends on unit-area yield of pasture grasses. Below the critical level of high yield, the more the yield, the more the number of pigs.

Secondly, the ratio also depends on the utilization of pond silt. The more pond silt is used, the fewer the number of pigs. Thirdly, it depends upon the quality of soil and pig manure and varieties of pasture grasses. On the basis of production test for ryegrass and sudan grass grown all the year round, the best allocation is 5–6 pigs per mu. The combination of fish-grass link has already been introduced in fish-cum-crop integration.

FISH FEEDS: CALCULATION STANDARD

In most cases, fish ponds in China are applied with whatever feeds available, a compound one. The formula of calculation is more complicated and it leads inconvenience to scientific management. In practice, a kind of feeds with common use and stable source often serves as a standard of calculation such as barley in southern part of Jiangsu Province or pelleted feeds in certain places. For green fodder, aquatic grass serves as a calculation standard. And then, on the basis of feed sources available and the needs of cultivated spp., the feed amount should be converted into the actual amount of feeds according to the equivalent ratio between actual amount and standard amount.

EX. A farm expects 1000 T barley, the department of goods and materials can only provide 50% of it, the rest are substituted by 20% bean cakes, 20% brans and 10% pellets. Evalution: the exact quantity of each substitutes.

Solve: According to the available informations, equivalent ratio between barley and the substitutes is 3/4 for bean cake, 2 for brans, ½ for pellets.

The quantity of bean cake = 1000 × 20% × 3/4 = 150 T.

The quantity of brans = 1000 × 20% × 2 = 400 T.

The quantity of pellets = 1000 × 10% × ½ = 50 T.

Manure for Fish Culture

Total requirements of manures are the sum of manures needed at different stages. If manures applied come from different sources, a kind of manures with common use and stable source serves as calculation standard. The actual amount can be calculated out according to equivalent ratio. The target amount of manure at a certain stage, M is:

$M = (Y_1 - n\ Y_2)\ C$

Where,

Y_1: the target net yield of filtering fish

Y_2: the target gross yield of feed-eaters

n: net yield of filtering fish which feed on the plankton produced out of excreta of the feed-eaters per unit gross yield.

However, the rate varies with the species and feeds and ecological factors. Generally, it ranges from 0.2 to 0.6;

C: the manure conversion coefficient of a certain kind of manures to fish flesh

In fact, this coefficient results from many ecological factors and will be obtained in the actual practice in each specific farm. The animal manure takes faeces as standard. The urine should be converted into faeces by certain ratio.

A certain farm is expected to produce herbivorous fish and filtering fish of 15000 kg each in grow-out pond with terrestrial grass to feed herbivorous fish and with livestock manure to fertilize pond water. Now there is about 100,000 kg of pure pig excreta. If the rest is replaced by cowdung, how much will it be needed?

Solve: Based on the experiment, the conversion coefficient of pig excreta and cowdung is 25 and 40 respectively. Suppose 0.5 kg of filtering fish can be raised by virtue of plankton produced by excreta of herbivorous fish. We use pig excreta as a standard.

$M = (15{,}000 - 0.5 \times 15000) \times 25 = 187500$ kg.

Area of Crop Production

Calculate grass planting area only because the economic returns of planting manure crop and fodder crop are poorer than that of planting grasses for fish farming.

The number of animals matched with fish farming and the construction area of animal shed.

WEB OF INTEGRATED MANAGEMENT

In order to raise the utilization efficiency of energy, to decrease cost, to increase products and income, several models could be interwoven to an integration web thru longitudinal and transverse nutrients cycles and energy flow in line with the local conditions. There are three kinds of integration web so far.

Multi-level Comprehensive Utilization Web

In the light of different utilization of nutrients by various specialized trader the trades can be connected in series with nutrient cycle and energy flow as a master line. An integration wed can be regarded as a number of connected nutrient cycles and energy flow. The popular form is recycling of animal manures.

That is to say, in recycling of chicken manure, the spilt and undigested chicken feeds can be fully utilized. Sgan Bei Livestock and Aquaculture Farm adopts this kind of integration web. Not only can chicken manure be fully utilized but also it can save about 25kg of fine feeds for pig. increasing net income by 47 yuan/pig. Furthermore, through simple deodorization and sterilization, chicken manure is mixed with fine feeds as composite pellets, which can be used to feed chicken again and then chicken manure can be fed to pig.

In Zhejiang Province, there is a common multi-level comprehensive utilization model, that is, cow-mushroom-earthworm-duck-fish or cow-mushroombiogas-earthworm-duck-fish. Cow dung can be used to grow mushroom; cow urine for fish farming; mushroom soil for earthworm culture; and then, earthworm is fed to ducks, the excreta of which is applied to fish ponds. The levels can be rearranged in line with the local conditions. Thus, different constituents of cow excreta are utilized at different levels or the new constituent produced at the preceding level can be utilized by the next level.

Parallel-cennected Web of Multi-trades

With fish farming as the centre, there may be a parallel connection of multi-trades, such as fish-livestock-crop integration to create sources of feeds and fertilizers. The products and wastes are utilized reciprocally among all trades. Most of integrated fish farms adopt this kind of integration web. E.g. in Xiang Yang Aquaculture Farm, fish farming is combined with pig, cow and geese raising.

Animal manures which is self-supporting are used for fish farming and for crops such as pasture grasses, squash, sweet potatoes and soybean and at the same time for cultivation of aquatic plants such as water hyacinth, Spirodella spp. and Wolffia arrihza which are used as feedstuff for fish, pig and geese. Although soybean is planted on pond dyke in the latter part of a year, water hyacinth can offer 80% of green fodder for fish and pig.

3. Fish-livestock-crop and agriculture-industry-commerce web

On the basis of parallel connection of fish-livestock-crop integration, all trades develop on both input and output ends. On input end, feedstuff processing industries can be set up and food can be utilized at multilevels; on the output end, the products of all trades needs processing and marketing. The waste of processing factories is used to raise livestock and poultry. This is a combination, which can raise the level of comprehensive utilization of natural resources and the rate of utilization of energy and can increase the products of fish, livestock and poultry resulting in the increment of value, income and job opportunity.

In 1970's, this type of integration web was set up in Dong Hu Fish Farm, Siang Yin County, Hunan Province . In recent years, the management structure and the scale of this web have developed, further, *e.g.* Helei Fish Farm has developed into an aquaculture, industry and commerce combination. On the axis of abscissae, they integrated fish farming with grass cultivation, pig, cow and duck.

On the axis of ordinates, they process crops, pupae so as to provide green fodder, pupae waste water for fish, livestock and poultry at the end of input; at the end of output, they process the products of livestock. *e.g.* making preserved egg, slaughtering ducks. The waste of processing factories can be used as feeds for fish, livestock and poultry or fertilizers for fish ponds. The processed products go to market home and abroad. Thus, the economic efficiency can be raised greatly. For example, in 1982, the average net profit was 6.78 Yuan/duck. The people who are engaged in duck raising and egg processing occupy 17% of the personnel of the farm while the net profit accounted for 43%.

APPRAISAL OF ECONOMIC RETURNS OF INTEGRATED FISH FARMING

In term of scope, investigation can be divided into general surveying, sampling and typical surveying. In terms of form, live coverage and questionaire through correspondence. Investigation through various methods can get better results, nevertheless, general surveying combining with typical surveying are commonly practised. Before investigation, the objective, contents and targets ought to be fixed; the precis of investigation be designed; the table of investigation be drawn up and then the surveying can be conducted deeply and widely as planned.

EXPERIMENT METHOD

Due to the weak points of investigation, the trial and error procedure should be performed for complement and verification. Experiment on the site should be held before any new decision on technical measurements is adopted. In order to give a correct appraisal, it's necessary to get technical and economic data as complete and accurate as possible.

Data Analysis and Processing

There are several methods of data analysis:

i. Comparative analysis
ii. Cut and try

iii. marginal analysis
iv. regressional analysis
v. linear analysis

Selecting the Best after the Appraisal

After the appraisal, the best plan can be selected in line with the local conditions and enterprise capabilities by using the abovementioned methods and overall & systemetical analysis.

Procedure

The procedure of appraisal of economic returns of integrated fish farming technique.

4

Challenges of Fisheries in Biotechnology

INTRODUCTION

In this chapter, biotechnology is defined as "any technique that uses living organisms to make or modify a product, to improve plants or animals, or to develop microorganisms for specific uses" (after the Office of Technology Assessment, US Congress). The document focuses on the development and application of modern biotechnology which is based on the use of new enabling techniques of recombinant-DNA technology, often referred to as "genetic engineering".

Biotechnology has great potential to influence and benefit livestock and fisheries. Modern techniques of biotechnology offer the potential of moving any cloned gene from any organism into any other organism and confer much greater precision and speed in achieving results as compared to conventional techniques. In conjunction with conventional technologies, modern biotechnology holds promise of increased and sustained productivity, efficient processing for improved product diversification and utilization, adaptation of product quality to functional requirements, and decreased reliance on agrochemicals and other external inputs.

It may also promote better conservation and use of genetic resources, and environmentally friendly management of natural resources. However, the number of marketable products and their influence at the farm level still seems to be limited, but is likely to increase in the next decade.

Biotechnology also poses certain challenges. These are largely determined by how, where and when it finds application. In general, the fast-paced research, predominantly funded by private-sector investment, and use of intellectual property rights in industrialized countries are seen as evidence that the application of biotechnology will hold the key to competitiveness and comparative advantage in many fields, including agriculture and food.

Biotechnology, with its vast potential and challenges, is thus of the utmost importance to agricultural development, and hence to FAO. The science of developing biotechnological tools cannot make a distinction between developing and developed countries. However, the application of such tools in the development process requires preconditions which are easily found in many developed countries, but which hardly exist in most developing countries. The current paths of research and development have given rise to concern that the disparity in harnessing biotechnologies for agricultural and economic development may increase between industrialized and developing countries.

The preamble to the FAO Constitution requires that the Organization work towards raising levels of nutrition and the standards of living of all peoples, securing improvements in the efficiency of the production and distribution of all food and agricultural products, improving the conditions of rural populations and thus contributing towards an expanding world economy. Therefore, one of the major tasks for the Organization is to ensure that the benefits of biotechnology will be shared by people in the north and the south, in both large and small, and rich and poor countries.

In recent years, various FAO bodies and conferences such as the Tenth Session of FAO's Committee on Agriculture, 1991; the FAO Conference 1989; Sessions of the FAO Commission on Plant Genetic Resources, 1989 and 1991; FAO/CTA International Symposium on Plant Biotechnologies for Developing Countries, 1989; and the 20th FAO Regional Conference for Asia and the Pacific, 1990, have strongly recommended that FAO, as the lead UN specialized agency for food and agriculture, must respond to the potential and challenges of modern biotechnology for agriculture, forestry and fisheries. This paper, therefore, examines the pattern of biotechnological development in developed and developing countries, the application, impacts and potential of biotechnologies in agriculture, forestry and fisheries. It then analyses some of the issues and concerns in biotechnology such as trade substitution, biosafety, and intellectual property rights, with special reference to developing countries. Finally, it elaborates the elements of FAO's policy and strategies on biotechnology.

DEVELOPMENTS IN BIOTECHNOLOGY

There exists a gradient of biotechnologies, depending on the degree of sophistication, complexity, stage of development and application. The lower side of the gradient comprises simpler but widely used techniques such as in vitro culture, rhizobium technology, fermentation. The upper side of the gradient includes advanced techniques involving genetic engineering. The gradient of biotechnologies may be matched with the gradient of national capabilities, economic investment and efforts to provide the possibility of choosing the appropriate techniques and approaches with the most positive impacts. The pattern of use and development of biotechnologies in developed and developing countries thus varies considerably.

THE PATTERN IN DEVELOPED COUNTRIES

In the industrialized countries, a new pattern of biotechnological research funding has emerged. With the availability of property right protection of biotechnologies and prospects of vast markets for biotechnology products and the techniques, the bulk of the research is funded, carried out, and controlled by the private sector. For instance, since 1976 when GENENTECH, the first biotechnology company, was established, the new techniques have spawned a variety of industries that now comprise more than 400 start-up firms, more than 200 established firms that have diversified into biotechnology, and more than 200 supply firms in the United States alone.

The new biotechnology industry in the United States produced pharmaceutical, diagnostic tests and agricultural products worth close to US$2 billion in 1990. A similar trend is seen in Europe and Japan. It is estimated that about 60 percent of the biotechnology research and development funding in industrialized countries is from the private sector. Thus, the private sector has been a major force in enhancing the capability of these countries in this field.

Research institutions in the public sector are now generally required to raise a substantial part of their budget from non-government sources, via contractual research, licensing agreements and royalties. This is tending to increase secrecy over research findings and to hinder free scientific communication. University professors, researchers and government institution scientists are increasingly becoming entrepreneurial and are entering private industry.

Another important trend is that large multinational corporations are purchasing smaller seed and biotechnology companies and diversifying their holdings. This allows them to develop a package sale of chemicals, seeds and equipment.

Heavy involvement of the private sector and market considerations greatly influence the topics and commodities chosen for research. Major crops, commodities and farming systems of great socio-economic importance to the developing world, but of little international market importance, do not figure in the biotechnology research agenda of industrialized countries. Furthermore, these countries are keen to reduce their production costs, increase the productivity, quality and value of their products and, thus, improve their overall competitiveness in the world market.

THE PATTERNS IN DEVELOPING COUNTRIES

Biotechnology facilities are being established in most developing countries. However, the level of research, development and use of biotechnology for agriculture, forestry and fisheries in the developing countries is generally far below the level in the industrialized countries. Among developing countries, the status varies considerably. A few, such as Brazil, Mexico, India, China and The Republic of Korea, have sought to gain full scientific and technological capacity, especially in agricultural biotechnology. Others, such as Indonesia, Malaysia, the Philippines and Thailand and a few countries of Latin America, have built the capacity to apply biotechniques and develop biotechnologies useful for their agriculture and food industries. The participation of the private sector in gaining biotechnological capacity is not significant in most of these countries.

Many developing countries have inadequate funding, poor human resources and limited access to information, resulting in a relatively low-level capacity for research and technology development and exploitation, especially in the field of modern biotechnology research, which is costly and requires highly trained personnel. Most developing countries are vague as to their immediate aims in agricultural biotechnology. Few have appropriate proprietary-right protection systems or mechanisms to increase their access to protected techniques and products. Furthermore, there is negligible involvement of the private sector, which accentuates the problem of insufficient attention to biotechnology.

One of the major constraints on biotechnological development in developing countries is the poor quality and extent of higher education in frontier sciences, especially molecular biology.

Besides, there are no or very weak links between universities and research institutions to reinforce each other and to utilize synergistically the scarce trained personnel resources. The university experts play a negligible role in national policy formulation, including that on biotechnology. Furthermore the universities are generally not oriented for efficiency in commercialization and are therefore not able to market biotechnology products of their own or others' inventions.

Some international and donor-country public sector institutions strive to ensure that attention is given to solving, through biotechnology, the food and agriculture problems of developing countries. Among this group are the International Agricultural Research Centres (IARCs) located or operating in developing countries. Capacity exists in several of these centres for utilization of advances in biotechnologies to solve certain problems of plant production and protection and of livestock production and health, and capacity is being built in others.

For example, the International Rice Research Institute (IRRI) is incorporating modern molecular biology in its rice research activities. Its programme is linked closely to the Rockefeller Foundation Biotechnology Network. The World Bank, the International Service for National Agricultural Research (ISNAR) and the Australian Government undertook a major study on the likely impact of modern biotechnology on agriculture. This study is being followed up by the participants by an expansion in the World Bank lending for biotechnology and by ISNAR's establishing an Intermediary Biotechnology Service to provide advisory services to national agricultural research systems.

APPLICATIONS, IMPACT AND POTENTIAL

Biotechnological research and development are moving at a very fast rate. For instance, three years ago, transgenics in rice, especially in japonica rices, were considered rather difficult because of the problem of regeneration through protoplast culture. But today, a large number of transgenics both in indica and japonica types are available and are being tested at various stages. It is therefore difficult to predict the potential and impact of a given technology beyond five years or so. This section gives a brief account of the current level of technology and refers to the developments likely to occur in the next three to five years.

The potential of genetic engineering in medicine has received much attention. The potential for advances in agriculture, forestry and fishery are similarly bright. Biotechniques are already being used to create new strains of crop plants, new plant and animal diagnostic products, animal vaccines, biological pesticides and herbicides, other biological control agents, and modifications in domestic animals used for food production.

In some cases, applications are being held in check by the need for still more research to ensure that there are no harmful effects (or, if there are, they are greatly outweighed by the benefits) and by the slow pace of evolution of regulations governing the release of genetically modified organisms or products for general use or acquisition.

Biotechnological research, especially genetic engineering, on problems of field and tree crops, forest species and fisheries has a relatively short history. The knowledge of biological aspects of species and ecosystems needs to be expanded in may cases before new biotechnologies can be applied effectively. Even where transfer of one or more genes can be achieved reliably, expression of the transferred gene(s) may not occur in the expected way. The search for and identification and cloning of useful genes, effective vector systems, methods of gene transfer and promoter mechanisms should be intensified.

A perspective on both the difficulties and intensity of the research effort suggests that diagnostic tools will speed solutions in the next five to ten years to certain disease problems in several crop, forest, animal and fish species, such as black sigatoka of banana and plantain, and virus/viroid diseases of coconut and rice.

Improved tolerance to certain physical stresses, such as heat tolerance in potatoes, may also be achieved, but prospects for improved resistance of varieties and clones to most abiotic stresses are longer term.

Livestock

Among agricultural and allied fields, animal production and health have benefited the most from biotechnology, although practical use of transgenic livestock is only a future possibility. Wide use of monoclonal antibodies for efficient diagnostics, leading to safe and specific treatments of animal diseases, is a major breakthrough.

Through genetic engineering, vaccines for the prevention of viral, bacterial, and parasitic animal diseases are rendered more effective and safe. Tailored vaccines exist for pig scours, chicken bursar disease and cattle tickborne diseases. Pathogen-specific vaccines are attractive goals. Other interesting possibilities are endocrine-directed vaccines to stimulate twinning in beef cattle, immunocastration, livestock growth rate increase and vaccines that compensate for various stress-induced production losses.

Advances in genetic engineering will also facilitate the production of male-only populations of screwworms, tsetse flies, ticks and various other ectoparasites for use in the sterile release technique of control or eradication. Furthermore mammalian tissue culture may replace whole animals in the 1990s for toxicity testing of certain chemicals. The culture technique can also be exploited for studying and analyzing pesticide metabolism and for herbicide pre-screening. In vitro fertilization and embryo sexing techniques have considerably increased the use of embryo transfer techniques for cattle breeding and trade. The value of the approach will be further enhanced if embryo cloning techniques can be reliably employed. Microbial and enzymatic treatment of roughages and genetic engineering of rumen bacteria both have great potential to improve animal nutrition. Growth hormones can be produced by genetically engineered microorganisms in quantities and at the low cost necessary for widespread use to speed and increase milk and lean meat production. Biotechnological tools (embryo culture, gene cloning, etc.) may also be used for conservation of genetic resources.

Forestry

Successful regeneration through micropropagation or somatic embryogenesis has been achieved for about 100 forest species although, for most, considerable development work would be required before commercial propagation could be contemplated. In breeding programmes incorporating clonal testing, inclusion of a micromultiplication phase may facilitate more rapid deployment of superior genotypes than that afforded by sequential multiplication by cuttings. This will particularly be the case when gene mappling techniques are sufficiently advanced to permit accurate identification of superior genotypes without replicated clonal testing in the field, especially within a breeding line where substantial genetic disequilibrium exists. Furthermore, gene mapping may allow eventual identification of valuable genes, including those that contribute to quantitative traits.

As regards the application of genetic engineering, many major traits of commercial importance to forestry are under polygenic control, and much remains to be learned about the operation of the genes involved before this technique can have a major impact for these characters. An earlier application of genetic engineering in some forestry operations may be the introduction of genes, some already available, known to confer insect of disease resistance. Other applications of biotechnology with obvious value for forestry, but as yet not widely supported by experimental successes, include: in vitro manipulation of the maturation state, *e.g.* the promotion of early flowering to reduce generation intervals, in vitro selection for traits such as disease resistance and tolerance to salinity; and the use of haploid cultures.

Fisheries

Two broad areas of biotechnology exist within the fisheries sector, namely natural products biotechnology (including mostly marine species); and aquaculture biotechnology. Commercially valuable products such as pharmaceutical compounds, pigments, oils, sterols, alginates and agarose are being extracted from micro-and macro-algae in many parts of the world. Aquatic invertebrates are currently being screened for biologically active compounds that may have, inter alia anti-carcinogenic or antiviral properties, whereas marine bacteria are currently utilized in treatments of oil spills and holding tanks in tanker ships.

Within aquaculture, induced increases in the chromosome complement (polyploidy) of commercially important species such as salmon and oysters have increased the growth and marketability of these species. Genetically engineered microbes can be used to produce fish growth hormones, which might then be used to improve feed conversion and growth rate. Synthetic reproductive hormones are produced commercially and used to regulate fecundity, breeding cycles, growth rates and sex determination in certain cultured species.

In attempts to increase desirable culture qualities such as growth rate, disease resistance, temperature tolerance and marketability, transgenic fishes containing introduced genes from other species have been produced on an experimental scale. The application of advanced biotechnology to the fisheries sector is a relatively recent practice. Therefore, continued research can be expected to reveal additional commercially viable applications and products within this sector.

PROCESSING AND PRODUCT QUALITY AND USES

Biotechnology could have great potential for improving the quality and diversifying the uses of agricultural, forestry and fisheries products. This technology could be used for processing products by both traditional and new methods in order to:

- improve the value, safety and quality of processed and unprocessed foods, such as genetically engineered tomatoes which soften more slowly and which could be processed at lower temperatures, resulting in a higher proportion of solids to liquid in canned tomatoes;
- enhance non-food uses of farm products and thereby increase their market value;
- develop new uses for farm products, especially those often producing surpluses, including energy production from biomass and recycling of wastes and by products. An example of such an application is found in ethanol production from cassava and sugar cane;
- produce algal cells from, for instance, Spirulina or Chlorella in mass culture for proteins;
- produce enzymes on a commercial scale, especially using the microbial system, as substitutes for enzymes isolated from animal or plant sources. For example, chymosin, the enzyme used in the manufacture of cheese, traditionally taken from the cow, can now be made from specially engineered microorganisms. Mass production of amino acids from molasses and other agricultural resources is another example. These amino acids can improve the nutritional value of food and animal feed;
- produce and use new probes for rapid detection and accurate identification of specific microbial pathogens in food, especially diagnostics for staphylococcal enterotoxins, and others; and
- enhance productivity through the application of newly developed processes such as the immobilized enzyme and plant tissue culture systems which are being used to produce useful secondary metabolites. Examples are diosgenin, a female contraceptive commonly used in China and India and shikonin, a natural pigment used for lipstick in Japan.

Biotechnology is being used in a number of countries to characterize the indigenous germplasm variability for therapeutic and biologically important substances. The use of biotechnology for the removal of toxic and antinutritional factors and for improving the quality of food products is gaining momentum.

These techniques include the elimination or reduction of cyanogenic compounds in cassava and the incorporation of high protein in potato and sweet potato.

CONSERVATION AND UTILIZATION OF GENETIC RESOURCES

New biotechnologies offer not only techniques to improve the conservation of genetic resources, but also new methods to identify, clone, transfer and express genes in different organisms. This has profound implications for the utilization of genetic resources, could broaden the germplasm base from which new genetic combinations can be created, and will allow scientists to pursue their breeding efforts with greater focus and speed.

In vitro conservation and the use of techniques for germplasm exchange of certain plant species are already proving considerably more efficient than conventional methods. Furthermore, the generation of gene libraries provides a valuable adjunct to the conventional germplasm conservation methods. Molecular methods such as Restriction Fragment Length Polymorph (RFLP), isozyme and protein analyses are already being used for the characterization of genetic resources and the identification of useful genes. However, massive use of crop varieties reproduced through in vitro culture, which contain genetically identical copies of the parent, could provoke increasing genetic erosion.

RESISTANCE/TOLERANCE TO STRESSES

Biotechnologies have been particularly effective in and could have great potential for developing genotypes resistant/tolerant to biotic and abiotic stresses commonly affecting crop, livestock, forest and fish species, and can therefore contribute to reducing inputs of pesticides and to stabilizing agriculture in marginal lands and inhospitable habitats. Genetically engineered de novo synthesis of biopesticides by host species for their own defence, development of biocapsules containing genetically engineered biopesticides and virus coat protein-mediated resistance represent tremendous opportunities for pest management.

It is anticipated that genetically engineered vegetable and cereal crops, resistant to certain insects and viruses or tolerant to herbicides, will be commercially available by 1993. Genetically engineered biopesticides are near approval for sale in some countries. The efficiency of insect-based control of pest arthropods, pathogens and weeds and the production of sterile males for insect control can be enhanced considerably through genetic engineering. Furthermore, biotechnologies are aiding conventional breeding in developing genotypes resistant/tolerant to high and low temperature regimes, drought and excess water conditions and saline and other problem soil conditions. A major concern about these solutions, however, is their sustainability if the improved attributes are based on one or only a few genes, as has often been the case.

SUSTAINABLE PRODUCTION

Biotechnology may be used as a tool in the sustained production of crops, livestock, forest and fisheries by providing opportunities to:

- develop genetically broad-based varieties/breeds resistant/tolerant to biotic and abiotic stresses;

- increase efficiency in the conservation and enhanced use of natural resources, including genetic resources and systems;
- reduce use of off-farm inputs such as pesticides;
- develop populations adapted to marginal land and problem soil conditions; and
- produce genetically engineered organisms for degradation of toxic wastes and detoxification of chemical residues on produce or land.

In order to be not only environmentally friendly but also socio-economically viable and attractive, a more vigorous and focused research and development approach, especially by the public sector, is necessary. The private sector should also be encouraged to achieve the desired goal of sustainable development through the use of sound biotechniques.

BIOTECHNOLOGY IN FISHERIES DEVELOPMENT

The gap between human demand and the availability of livestock products is huge in Africa, and imports have been increasing fast. The productivity of African breeds is very low: mature beef cattle of 4-5 years hardly weight 300 kg (as against 400 kg at one year for exotic animals); the best African cows produced 300 litre/lactation (as against 5 000 litres and more in Europe); 100 African ewes produced 50 lambs every two or three years (as against 150 lambs a year in developed countries).

Retrospectively, production policies that aimed at developing livestock through cross-breeding of exotic grade cattle with indigenous ones have generally failed. The new approach is not to make selection of local performant animals. Biotechnology offers good methods both for selection and cross-breeding schemes, essentially through Multiovulation Embryo Transfer (MOET) and the field diagnosis of pregnancy. Yet, there is wide diversity among the countries to use new technologies. For instance, even if simpler techniques such as Artificial Insemination (AI) are considered, it is seen that in East Southern Africa the technique is widely used, whereas in West and Central Africa, because of the predominance of nomadic transhumant livestock production, the technique is least developed.

MOET is well developed in few countries such as Zimbabwe and South Africa. Cattle farmers in these countries import frozen embryos from overseas. MOET is promising especially in the development of dairy cattle farms around large cities. Both AI and MOET are being used in Open Nucleus Breeding Systems (ONBS) to improve animal production. Small-scale ONBS projects are being carried out by FAO in small ruminants in Ghana and the Gambia. Moreover, by using semen from bulls, with genetic resistance to high temperature, diseases and insects common to African countries, embryos could be produced from superior cows of other regions and then implanted into surrogate mothers within the environment where improved livestock are needed. In animal feeding, biodegradation of low-quality forage is also a promising technology. It can be associated with chemical degradation techniques (use of urea and ammonia) to improve animal feeds during dry seasons.

In some of the countries, farm diagnosis of pregnancy is also being used by field workers by using a kit based on the difference of levels of progesterone in cow's milk.

In the field of animal health, Sterile Insect Technology (SIT) is well known in Africa. Sterile males of glossina are being reared in Bobo-Dioulasso and Burkina Faso for field application. Successful field projects in tsetse control have been carried out in Nigeria (BICOT Project) and in Burkina Faso. Newworld screwworm was eradicated in the Libyan Arab Jamahiriya in 199091 by the use of SIT.

In disease diagnosis, ELISA is being used to evaluate rinderpest immunity in vaccinated animals. Monoclonal Antibody Techniques (MAB) is not commonly used. The two techniques are very promising because they are very simple and most effective. Trials are under way to use the ELISA test in trypanosomiasis diagnosis. In vaccine production recombinant rinderpest vaccine trials done in the United States seem ready to be utilized in a mass vaccination campaign against rinderpest in Africa. There is an increasing level of activity in hybridoma work in sub-Saharan countries.

The work of ILRAD in Kenya is addressing the problem of diseases and parasites that affect cattle. Some of the common cattle diseases are East Coast Fever, tickborne diseases and trypanosomiasis. The focused approaches to researching on these diseases is expanding to yield results that will significantly benefit cattle breeders. The development of rinderpest vaccines remaining stable at ordinary temperatures was of high practical value.

Biotechnology in fisheries is recent in Africa and it is being applied essentially in private fisheries sector. Work is mostly confined to hormonal treatments for sex reversal and pawning, detection of fish and shrimps disease and development of fish feed.

BIOTECHNOLOGY IN FORESTRY

Forestry research in Africa, let alone forestry biotechnologies, is not well developed. Some success stories were experienced with rhizobium technology such as coniferous plantations in East and southern Africa and Casuarina plantations in Senegal. However, in recent years, there is a tendency to move towards genetic improvement and in. vitro culture. African forestry research institutions are actively engaged in laying the foundations for genetic improvement research. This is a very necessary prerequisite for biotechnology or any other sophisticated research to be effective.

According to a survey carried out by the FAO Regional Office for Africa (1987), 20 countries in the region, including the North African subregion, had established national forest seed centres, and FAO is helping, through a cooperative programme with France, to establish a seed centre in each Sahelian country. The survey further revealed that about 100 researchers in 22 countries are undertaking some kind of genetic improvement research, but only three countries are engaged in in vitro culture, with considerable inputs from foreign institutions and expertise in two cases. However, most of the research institutions are engaged in research on vegetative propagation for the establishment of clonal plantations.

BIOTECHNOLOGY IN SELECTED AFRICAN COUNTRIES

Summarized below is the development and application of biotechnology in ten selected African countries, representing the various subregions. The information is adopted from Bialy (1993), Chatsanga (1993), Komen and Persley (1993) and Sasson (1993).

BURUNDI

Biotechnology work in Burundi was confined to in vitro culture and the promotion of biological nitrogen fixation. In vitro micropropagation was performed for rice, maize and sorghum. At the University of Burundi, Bujumbura, yam, cassava and potato were propagated in the same way. Similar work was carried out at the Agricultural Sciences Institute of Burundi. A national programme aimed at improving potato, and propagating selected varieties on a large scale, was supported by the International Potato Centre (CIP). Potato plantlets were marketed by a private company at the same time as agricultural inputs. Demand for potato plantles was high and such a marketing and distribution mechanisms was appreciated by the partners involved.

At the Agricultural and Zootechnical Institute (IRAZ) of Gitega, serving Burundi, Rwanda and Zaire, banana was micropropagated. Banana plantlets were mainly sold to Rwanda, which had a small industry for processing banana beer and wine. Since IRAZ could not meet the demand for technical reasons, Rwanda created its own micropropagation production unit and the Institute was considering giving its micropropagation work to a private company, so as to concentrate on upstream research.

CAMEROON

Micropropagation techniques have been introduced for root and tuber crops in Cameroon. Development work at the Root and Tissue Research Project Biotechnology Laboratory has focused on mass-producing disease-free seedlings from the roots and tubers of cocoyam and yam. Embryonic tissue material has been used to produce cassava seedlings. The Cameroon biotechnology capability is enhanced by other projects under way at the

University of Yaoundé Biotechnology Centre which is using techniques such as cell and tissue culture, hybridoma development and recombinant DNA. The Centre's work on developing diagnostic antigens from Onchocerca volvulus, the causative agents for river blindness, is well recognized.

Bio/Afrique, a diagnostic biotechnology company in Yaoundé, was formed as a profit-making joint venture between the government of Cameroon, through the biotechnology centre at the University of Yaoundé, and private Cameroonian investors as equity partners; Applied Biotechnology (San Diego, CA) as a source of proprietary manufacturing technologies; Chiron, Crop. (Emeryville, CA), Tanox Biosystems (Houston, TX), New England Biolabs (Beverly, MA) and a number of European and American academic laboratories as scientific partners. The impact of this joint venture is yet to be felt.

Côte d'Ivoire

The Institut international de recherche scientifique pour le développement en Afrique (IIRSDA) is the leading centre for biotechnology development in the country. Yam, a very important food crop in West Africa, is being researched in order to accelerate its rate of propagation. Studies include genome characterization, disease-free micropropagation, karyotyping, protoplast fusions and in vitro conservation. Techniques of micropropagation of cells obtained from the calli of young leaves and flowers of oil palm are being standardized. Somatic hybridization and field trials of vitro plants are being pursued. The country is the world's number one producer of cocoa and the crop accounts for almost 50 percent of export earnings. Plans to use biotechnology to develop disease-resistant and superior quality cocoa are receiving due attention.

Ghana

Ghana was once the world's leading cocoa producer. By the 1970s, it ranked far behind Côte d'Ivoire, the new leader, Brazil and Malaysia. It nevertheless maintained a good reputation for cocoa, thanks to rigorous quality control. At the faculty of Agriculture of the University of Ghana, disease-resistant and higher-yielding cocoa plantlets were being multiplied through clonal propagation; thermotherapy was applied to eliminate viral and bacterial diseases. At the National Agricultural Research Institute, work was being conducted on in vitro conservation of various indigenous crop species. The Department of Veterinary Services of the University of Ghana was developing vaccines against some of the prevalent cattle diseases.

The Overseas Development Administration (ODA, United Kingdom) contributed £7.5 million of financial and technical assistance over four years to Ghana's Cocoa Research Institute and Cocoa Marketing Board. Technical assistance covered the provision of staff at the Institute and training of Ghanaians in the United Kingdom. ODA also supplied equipment for the rehabilitation of laboratories, in line with research priorities. Among other things, ODA provided assistance for the development of nucleic acid probes and Restriction fragment Lenght Polymorphism (RFLP) markers intended to understand better and prevent cocoa viral and fungal diseases. Research in these areas was conducted in collaboration with the John Innes Institute at Norwich.

Kenya

The Kenya Agricultural Research Institute (KARI) is pioneering the development of modern biotechnological approaches to existing programmes, especially tissue culture technology for producing uniform and disease-free seedlings of potato, pyrethrum and tea. Efforts are under way to develop inoculants for leguminous plants and transfer of the technology to farmers. Until 1993 the focal point for biotechnology coordination in Kenya was the National Advisory Committee on Biotechnology Advances and their Applications (NACBAA). It was composed of the directors of research institutes under the Ministry of Research, Science and Technology (covering agriculture, industry, health and the environment), plus representatives from the private sector. The Committee had responsibility for:

- developing a biotechnology strategy up to the year 2000;
- assessing the expected degree of scientific advancement;
- advising on modalities for handling biotechnology breakthroughs; and
- advising on the feasibility of a biotechnology park.

An outline for a national biotechnology programme was prepared in 1991 by NACBAA, identifying national priority areas, the development of regulatory guidelines, and the formation of a decentralized National Biotechnology Enterprise Programme. The programme would involve the creation of a National Biotechnology Education Centre, a Biotechnology Enhancement Fund, and a National Commission for Biotechnology to succeed NACBAA, which was disbanded after completing its task. The faculties of science and agriculture at the University of Nairobi and at Jomo Kenyatta University are also building expertise in this field, and the former intends to establish a biotechnology institute to provide a focus for biotechnology research and education.

Mauritius

The University of Mauritius identified three areas of biotechnology research to start with, namely upgrading of sugarcane byproducts, biological nitrogen fixation and plant tissue and cell culture. In collaboration with Imperial Chemical Industries (ICI) Ltd and using ICI airlift fermenter and continuous culture techniques, molasses were used as a substrate to produce yeast single-cell protein (microbial biomass); in association with Mauritius

Farms Ltd, trials were conducted on the use of this yeast biomass as a protein substitute in chicken feeds.

Investigations on the use of sugar-cane juice instead of molasses, improvement of the culture medium, selection of yeast strains, pH and temperature optimization, flocculation and sedimentation processes, yeast nutrient value and potential use as food and feed, were also planned. The digestibility of bagasse by ruminants and its nutritive value after a variety of biotechnological treatments to decrease its relative lignin content, followed by biodegradation, constituted another area of investigation. The University also foresaw the need for a databank on genetic engineering.

Several grain legume species were selected for pulse production locally to cut down on imports: Vigna mungo, Phaseolus vulgaris, Lablab purpureus, Glycine max, Lens culinaris and Cicer arietinum. Research work was being carried out on the selection of effective and persistent strains of Rhizobium for these legume species to be grown under various soil conditions and fertilizer. As regards other crops, after screening selected species of economic importance to Mauritius (tea, coffee, vanilla, fruit crops, Anthurium andreanum species and other flower species), their suitability to be cloned successfully through tissue culture was studied; these techniques were also expected to be used on a routine basis to produce selected plants with desirable attributes.

Among the requirements for biotechnology development in Mauritius, the following were identified: (i) a national programme for biotechnology development; (ii) national machinery to review advances in biotechnology applications in the Mauritian context, safety standards, training requirements, etc., taking into account assistance schemes of the United Nations system, the EEC and bilateral cooperation; (iii) a documentation centre; and (iv) funding arrangements and distribution of tasks.

Among the R&D projects, the following were considered as priority: fermentation processes, including anaerobic digestion of wastes and methane production, alcohol production, composting, degradation of lignocellulosic biomass; biofertilizers and biopesticides; virus-free crops, fruits and vegetables; production of hybrid horticultural and ornamental crops; dwarf tree crop varieties and fast-growing "energy" crops; soil-acidity tolerant crop varieties; microbial biomass for animal feed; embryo transfer in cows to increase milk yield; bovine and porcine growth hormones; diagnostic kits for common plant and animal diseases; vaccines for domestic animals (*e.g.* coccidiosis, Newcastle virus disease); and mass culture of micro-algae. An active effort is being made to link the public and private sectors.

Nigeria

Nearly ten government research institutes of the 24 in 1993 were involved in some kind of biotechnology research, largely related to agriculture and food processing. For instance, more palatable and sweeter varieties of cassava and yam were being developed; biotechnologies were applied to improve the shelf-life of palm wine and other foods; research was being carried out on the selection of wheat lines capable of withstanding the humid conditions in southern parts of the country.

Two institutes were able to apply advanced techniques, in addition to IITA in Ibadan: the Biotechnology Unit of the Institute for Agricultural Research and Training (Ibadan), which applied tissue culture to maize (mainly for yield improvement), cowpea and groundnut; and the Biotechnology Centre of the Anambra State University of Technology, focusing on food fermentation and industrial enzyme production.

The Nigerian Institute for Oil Palm Research (Benin), the Cocoa Research Institute of Nigeria (Ibadan), and the National Root Crop Research Institute (Umudike) applied plant tissue culture and other simpler biotechniques. The Oil Palm Institute is also working on the fermentation of oil palm sap to produce palm wine. ICRIN has also been concentrating on somatic embryogenesis, in vitro conservation of germplasm and the identification of microorganisms that could be used to improve the biochemical development of cocoa flavour during the fermentation process as well as those required for the digestion of pod husk on a large scale, to render it suitable as an animal feed.

FAO, the United Nations Development Programme (UNDP) and the International Plant Genetic Resources Institute (IPGRI) are supporting the National Centre for Genetic Resources and Biotechnology (CGRB), located at the Moor Plantation, Ibadan. It coordinates activities carried out in nine satellite laboratories, each of which has been given a mandate for research into specific biotechnology areas. Furthermore, the Ministry of Science and Technology (MOST) had set up a national committee on biotechnology and genetic engineering, with a view to pooling the scattered human resources in microbiology, genetics and molecular biology and to advising the Ministry and CGRB.

The research institutes involved in animal health were the National Livestock Research Institute, the National Animal Production and Research Institute (NAPRI), Shika, Zaria, and the National Veterinary Research Institute (NVRI), Vom, as well as the Nigerian universities at Sokoto, Ibadan, Maiduguri and Zaria, carrying out some biotechnology research. The major investigation areas included the development both of new breeds of cattle for beef and milk production, and of vaccines against a wide range of diseases.

Small-scale food processing was an important industry in Nigeria. Many products were the results of the simple but effective use of traditional biotechnologies, *e.g.* the production of cassava-based foodstuffs, palm wine and various other wines from local fruits such as mango, pineapple and banana, and lager beer from malted sorghum and maize.

In the early 1980s, researchers at the University of Ife identified a protein sweetener in the exceedingly sweet berry of a plant, Thaumatococcus danielli, common to the forests of southwestern Nigeria. It was subsequently confirmed that thaumatin, the sweetener in the tropical berry, was weight-for-weight at least 1 600 times sweeter than sucrose, and yet nutritious. The gene for thaumatin has since been cloned, and the protein sweetener was selling for more than US$ 5 000 per kilogram.

The Nigerian Government allocated about 45 percent of its total resources for biotechnology to agricultural biotechnologies, including those to be applied to mitigate post-harvest losses. Inadequate funding by the government and private corporations, as well as the lack of interest showen by multinational corporations was proving a major hindrance to biotechnology research and development.

The total funding available for all research coordinated by the National Centre for Genetic Resources and Biotechnology was less than US$ 250 000 in 1987, while IITA was spending more than US$ 2 million on its root crop investigations alone. Another bottleneck was the shortage of trained personnel.

Senegal

Work has mostly been concentrated on biological nitrogen fixation. At the Senegalese Institute for Agricultural Research (ISRA) and the French Institute of Scientific Research for Cooperative Development (ORSTOM) laboratory in Dakar, research carried out on actinorhizian microorganisms (Frankia) aimed to identify the most effective symbiotic associations with Casuarina and Allocasuarina genotypes, micropropagated in vitro.

The identified associations could be used in reforestation schemes to be implemented in Sahelian countries, and also in North Africa and the Far East. Similar work has been done on Acacia species and their symbiotic microorganisms (rhizobium).

At the same laboratory, Azolla species were studied with a view to being used as biofertilizer in paddy fields in West Africa, because of their high nitrogen-fixing capacity. Manipulation of the rhizobia of Sesbania rostrata which formed dense cover during the wet season in flooded zones was improved.

Field experiments carried out in the Casamance region using Sesbania rostrata as a green manure in rice cultivation showed that the paddy yield could be increased from 1.7 to 4 tonnes per hectare without any nitrogen fertilizer input. With a good field inoculation and good cultivation practices, this shrub legume could be used systematically and intensively in hydromorphic soils or in paddy fields.

The work should further be intensified and results shared with other ricegrowing countries in the region and outside. Other ISRA programmes focused on the biotechnological manipulation of nitrogen fixation in groundnut and Vigna unguiculata. A West African Regional Plant Biotechnology Laboratory, through financial assistance from France, was inaugurated in 1991, at the ISRA/ORSTOM centre. A staff of about 20 Senegalese and European researchers was expected to work in the laboratory, equipped for plant tissue culture and micropropagation. This laboratory has recently been designated one of the cooperating centres of the International Centre of Genetic Engineering and Biotechnology (ICGEB).

Zambia

The use of legume inoculants has been increasing rapidly in Zambia, demanding intensive research and development work in this field. Established in 1984, the Mt. Makulu Research Station Inoculum Production Facility has grown rapidly and promoted the application of biofertilizers over inorganic nitrogen usage.

An estimated 8 000 hectares of soybean are being inoculated annually with appropriate Rhizobium strains. Two new bioreactors were assembled and tested in 1991 at the Facility, each with a capacity to produce 100 litres of broth culture. The average annual production increase in legume inoculants on a peat-base carrier for several species has been 10-15 percent since 1987. A total of 75 000 packets, each 250 g, was supplied during 1991-1992.

Zimbabwe

The prime agency for coordinating biotechnology in Zimbabwe is the Research Council of Zimbabwe (RCZ). Its overall function is to advise the government on issues of science and technology. In its latest National Science and Technology Policy Statement, agricultural biotechnology research is a principal topic. The major trust of the biotechnology programme proposed by RCZ is to develop high-yielding crops, food technology, improved horticultural crops and improved methods of animal breeding, as well as to create a national gene bank. The proposed national programme and related issues were discussed at three workshops held in 1991.

The biotechnology programme in Zimbabwe has been spearheaded by scientists at the University of Zimbabwe which now offers an M.Sc. degree in biotechnology. The most applied early work was with the tissue culture of potato and tobacco. A private company has now taken over commercial micropropagation of potato and strawberries. Improved production of rhizobium inoculants is being contemplated.

The Bioteehnology Forum, initially supported by the Netherlands, leads the national planning in biotechnology. A Biotechnology Research Institute, emphasizing applied biotechnology work, has been established. A number of projects utilizing recombinant DNA technology are under way. These include the development of a Salmonella probe, cloning of the S gene of the Hepatitis B virus, characterization of the cowpea mosaic virus, milk fermenting local microorganisms, and immobilized enzymes. A number of Ph.D. students are involved in these projects.

FUTURE PROSPECTS AND STRATEGIES

Alleviating the widespread food gap and attaining food security or food reliance and nutritional adequacy should be the highest national priority in almost all African countries. In order to achieve this foremost goal, production, productivity and sustainability of food and agriculture must be greatly enhanced through the generation, development and transfer of appropriate technologies. While the potential of currently available technologies should be rationally harnessed, new and emerging technologies such as biotechnologies, should be used in conjunction with, not as substitute for, existing conventional technologies.

Biotechnology is offering innovative possibilities for increasing crops, livestock, forestry and fisheries production and for protecting the environment by the reduced use of agrochemicals, conservation of genetic resources, bioremediation and recycling of wastes. In vitro culture techniques for micropropagation, freeing planting materials of viruses and other pathogens, the production of secondary metabolites and germplasm conservation are being widely used.

Genetic mapping of major tropical crops as an aid to conventional plant breeding programmes; plant virus resistance by genetic engineering of the host plant; novel biocontrol agents for pest control to reduce pesticide use; and the use of efficient and new biofertilizing agents are now being increasingly used mostly in developed and in a few developing countries. Embryo transfer techniques and the use of diagnostic kits for detection of diseases both in plants and animals, and the use of highly specific and effective vaccines are proving of great value to animal production and health.

Biotechnologically designed new feeds, hormones and disease control agents are already favourably impacting fish production. But, in Africa, these technologies and products are being used only rarely, in a limited way and in an ad hoc manner. The slow and poor development of modern biotechnology in Africa may be ascribed to: (i) lack of appropriate policies, strategies, plans and programmes and the absence of political will in general; (ii) inadequacy of human resources, paucity of financial investments, poor research and laboratory infrastructure, absence of priority-setting mechanisms and insufficiency of mechanisms to analyse global market and technology trends; and (iii) negligible involvement of the private sector and absence of effective regulatory measures.

In view of the above, the African strategy for biotechnological development should include both the provision of direct support for biotechnological research, product development, technology assessment and transfer and the creation and fostering of a climate where biotechnology can develop in a balanced manner. This will largely depend on governmental and public awareness, political will, policies and programmes for building national capabilities and strengthening/developing regional and international cooperation.

Creative partnerships between public and private sector interests are critical in establishing a competitive strategy in biotechnology. While developing institutional mechanisms best suited to a given country, it is essential to establish effective interdisciplinary collaboration among molecular biologists, breeders, agronomists and plant protection specialists and to forge and menage linkages among researchers, extensionists, farmers and other users of bioproducts and techniques. It may be generally preferable to build the new initiatives on traditional strengths in agricultural, forestry and fishery research, rather than creating totally new institutes. It should be kept in mind that biotechnology in crop and livestock production will be applied through the creation of new genotypes possessing novel characteristics, and this is possible only when conventional breeding and production systems are already in place.

Development of trained personnel should be a high priority in Africa. The important areas of training that are required for a successful biotechnology programme are microbiology, plant biology, zoology, cell biology, virology, genetics, biochemistry and cell and molecular biology.

In most of sub-Saharan Africa, the people trained in these areas are in very short supply. Even biology and tissue culture training is not in the currriculum of most tertiary education institutions. Very few universities offer training with adequate laboratory facilities in molecular biology and recombinant DNA technology.

Those that are trying to offer such training are constrained by the lack of equipment and staff. In recent years, however, some donor-supported formal training programmes in biotechnology have been developed. For instance, the M.Sc. in biotechnology at the University of Zimbabwe was assisted by SAREC (Sweden) and DGIS/Ministry of Foreign Affairs (the Netherlands).

The donors provided the funding for both the teaching/research equipment as well as funds to buy supplies and to cover the cost of bringing Dutch and Swedish biotechnology specialists to Zimbabwe to give special lectures in areas in which there was no local competence. Students from Zimbabwe have also been admitted to universities in the donor countries for Ph.D. degrees. Such arrangements exist and/or are being developed in several other African countries.

The actual and potential effects of biotechnology development are not always positive. In the African context, the negative effect is the displacement of crops at present grown for export in Africa by biotechnologically designed "industrially" produced analogous in industrialized countries.

African countries should be aware of such developments and diversify their products or increase quality and cost-effectiveness of their products not only to maintain the international market share but also to enhance or open new opportunities.

FAO and other concerned international systems should assist African countries in developing early warning systems whereby the potential relative substitution effects could be monitored and strategic adjustments brought about in time to avert damage.

In industrialized countries, where biotechnology is essentially in the hands of the private sector, the work is focused on areas and commodities that will fetch higher economic returns to often costly investments in biotechnology research and development.

Commodities such as bananas plantations, cocoa, oil palm, roots and tubers, which are of high socio-economic importance to Africa, should be on priority biotechnology research agendas of countries in the region. International agencies and research centres should also pay due attention to "orphan" commodities.

Most of the African countries with an interest in modern biotechnology are aware of the need to institute adequate biosafety provisions and establish appropriate intellectual property and patenting systems.

The GATT agreements have brought these issues to the freefront, and they must be attended to in time to enable countries to access global markets. Preliminary work on developing biosafety guidelines is underway in many African countries.

The status of the issue of intellectual property rights in the region is still in a state of flux. Policymakers and biotechnologists in the region, in consultation with concerned international agencies, need to address this issue more decisively. Access to information by establishing the necessary national and regional databases and information systems should be strengthened to render priority setting, technology assessment and transfer mechanisms more effective.

BACTERIOLOGICAL CHANGES

The Bacterial Flora on Live Fish

Microorganisms are found on all the outer surfaces (skin and gills) and in the intestines of live and newly caught fish. The total number of organisms vary enormously and Liston (1980) states a normal range of 10^2–10^7 cfu (colony forming units)/cm^2 on the skin surface. The gills and the intestines both contain between 10^3 and 10^9 cfu/g .

The bacterial flora on newly-caught fish depends on the environment in which it is caught rather than on the fish species . Fish caught in very cold, clean waters carry the lower numbers whereas fish caught in warm waters have slightly higher counts. Very high numbers, *i.e.*, 10^7 cfu/cm^2 are found on fish from polluted warm waters. Many different bacterial species can be found on the fish surfaces. The bacteria on temperate water fish are all classified according to their growth temperature range as either psychrotrophs or psychrophiles. Psychrotrophs (cold-tolerant) are bacteria capable of growth at 0°C but with optimum around 25°C.

Psychrophiles (cold-loving) are bacteria with maximum growth temperature around 20°C and optimum temperature at 15°C . In warmer waters, higher numbers of mesophiles can be isolated.

The microflora on temperate water fish is dominated by psychrotrophic Gram-negative rodshaped bacteria belonging to the genera *Pseudomonas, Moraxella, Acinetobacter, Shewanella* and *Flavobacterium.*

Members of the *Vibrionaceae (Vibrio* and *Photobacterium)* and the *Aeromonadaceae (Aeromonas* spp.) are also common aquatic bacteria and typical of the fish flora. Gram-positive organisms as *Bacillus, Micrococcus, Clostridium, Lactobacillus* and coryneforms can also be found in varying proportions, but in general, Gram-negative bacteria dominate the microflora. Shewan (1977) concluded that Gram-positive *Bacillus* and Micrococcus dominate on fish from tropical waters. However, this conclusion has later been challenged by several studies which have found that the microflora on tropical fish species is very similar to the flora on temperate species. A microflora consisting of *Pseudomonas, Acinetobacter, Moraxella* and *Vibrio* has been found on newly-caught fish in several Indian studies . Several authors conclude, as Liston (1980), that the microflora on tropical fish often carry a slightly higher load of Gram-positives and enteric bacteria but otherwise is similar to the flora on temperate-water fish.

Aeromonas spp. are typical of freshwater fish, whereas a number of bacteria require sodium for growth and are thus typical of marine waters. These include *Vibrio, Photobacterium* and *Shewanella.* However, although *Shewanella putrefaciens* is characterized as sodium-requiring, strains of *S. putrefaciens* can also be isolated from freshwater environments. Although *S. putrefaciens* has been isolated from tropical freshwaters, it is not important in the spoilage of freshwater fish.

Table: Bacterial flora on fish caught in clean, unpolluted waters

Gram-negative	Gram-positive	Comments
Pseudomonas	*Bacillus*	
Moraxella	*Clostridium*	
Acinetobacter	*Micrococcus*	
Shewanella putrefaciens	*Lactobacillus*	
Flavobacterium	Coryneforms	
Cytophaga		
Vibrio Photobacterium		*Vibrio* and *Photobacterium*
Aeromonas		are typical of marine waters; *Aeromonas* is typical of freshwater

In polluted waters, high numbers of *Enterobacteriaceae* may be found. In clean temperate waters, these organisms disappear rapidly, but it has been shown that *Escherichia coli* and *Salmonella* can survive for very long periods in tropical waters and once introduced may almost become indigenous to the environment . The taxonomy of S. *putrefaciens* has been rather confused. The organism was originally associated with the *Achromobacter* group but was later placed in the Shewan *Pseudomonas* group IV.

Based on percentage of guanine+ cytosine (GC%) it was transferred to the genus *Alteromonas,* but on the basis of 5SRNA homology it was reclassified to a new genus, *Shewanella*. It has recently been suggested that the genus *Aeromonas* spp. which was a member of the *Vibrionaceae* family be transferred to its own family, the *Aeromonadaceae* . Japanese studies have shown very high numbers of microorganisms in the gastrointestinal tract of fish, and as such numbers are much higher than in the surrounding water, this indicates the presence of a favourable ecological niche for the microorganisms.

Similarly, Larsen *et al.* (1978) reported up to 10^7 cfu/g of vibrio-like organisms in the intestinal tract of cod and Westerdahl *et al.* (1991) also isolated high numbers of vibrio-like organisms from the intestines of turbot. *Photobacterium phosphoreum* which can be isolated from the surface can also be isolated in high numbers from the intestinal tract of some fish species. On the contrary, some authors believe that the microflora of the gastrointestinal tract is merely a reflection of the environment and the food intake.

Microbial Invasion

The flesh of healthy live or newly-caught fish is sterile as the immune system of the fish prevents the bacteria from growing in the flesh. When the fish dies, the immune system collapses and bacteria are allowed to proliferate freely. On the skin surface, the bacteria to a large extent colonize the scale pockets. During storage, they invade the flesh by moving between the muscle fibres. Murray and Shewan (1979) found that only a very limited number of bacteria invaded the flesh during iced storage.

Ruskol and Bendsen (1992) showed that bacteria can be detected by microscope in the flesh when the number of organisms on the skin surface increases above 10^6 cfu/cm^2. This was seen at both iced and ambient temperatures. No difference was found in the invasive patterns of specific spoilage bacteria *(e.g.,* S. *putrefaciens)* and non-spoilage bacteria.

Since only a limited number of organisms actually invade the flesh and microbial growth mainly takes place at the surface, spoilage is probably to a large extent a consequence of bacterial enzymes diffusing into the flesh and nutrients diffusing to the outside.

Fish spoil at very different rates, and differences in surface properties of fish have been proposed to explain this. Skins of fish have very different textures. Thus whiting *(Merlangius merlangus)* and cod *(Gadus morhua)* which have a very fragile integument spoil rapidly compared to several flatfish such as plaice that has a very robust dermis and epidermis. Furthermore, the latter group has a very thick slime layer, which includes several antibacterial components, such as antibodies, complement and bacteriolytic enzymes.

SPECIFIC SPOILAGE ORGANISMS

Bacteria on fish caught in temperate waters will enter the exponential growth phase almost immediately after the fish have died. This is also true when the fish are iced, probably because the microflora is already adapted to the chill temperatures. During ice storage, the bacteria will grow with a doubling time of approximately 1 day and will, after 2-3 weeks, reach numbers of 10^8-10^9 cfu/g flesh or cm^2 skin. During ambient storage, a slightly lower level of 10^7-10^8 cfu/g is reached in 24 hours. The bacteria on fish caught in tropical waters will often pass through a lag-phase of 1-2 weeks if the fish are stored in ice, whereafter exponential growth begins. At spoilage, the bacterial level on tropical fish is similar to the levels found on temperate fish species.

If iced fish are stored under anaerobic conditions or if stored in CO_2 containing atmosphere, the number of the normal psychrotrophic bacteria such as S. *putrefaciens* and *Pseudomonas* is often much lower, *i.e.*, 10^6-10^7 cfu/g than on the aerobically stored fish. However, the level of bacteria of psychrophilic character such as P. *phosphoreum* reaches a level of 10^7-10^8 cfu/g when the fish spoil.

The composition of the microflora also changes quite dramatically during storage. Thus, under aerobic iced storage, the flora is composed almost exclusively of *Pseudomonas* spp. and S. *putrefaciens* after 1-2 weeks. This is believed to be due to their relatively short generation time at chill temperatures and is true for all studies carried out whether on tropical or temperate-water fish. At ambient temperature (25°C), the microflora at the point of spoilage is dominated by mesophilic *Vibrionaceae* and, particularly if the fish are caught in polluted waters, *Enterobacteriaceae.*

A clear distinction should be made between the terms spoilage flora and spoilage bacteria since the first describes merely the bacteria present on the fish when it spoils whereas the latter is the specific group that produce the off-odours and off-flavours associated with spoilage. A large part of the bacteria present on the spoiled fish have played no role whatever in the spoilage. Each fish product will have its own specific spoilage bacteria and the number of these will, as opposed to the total number, be related to the shelf life.

It is not an easy task to determine which of the bacteria isolated from the spoiled fish are those causing spoilage, and it requires extensive sensory, microbiological and chemical studies. First, the sensory, microbiological and chemical changes during storage must be studied and quantified, including a determination of the level of a given chemical compound that correlates with spoilage (the chemical spoilage indicator). Second, bacteria are isolated at the point of sensory rejection. Pure and mixed cultures of bacteria are screened in sterile fish substrates for their spoilage potential, *i.e.*, their ability to produce sensory (off-odours) and chemical changes typical of the spoiling product. Finally, the selected strains are tested to evaluate their spoilage activity, *i.e.*, if their growth rate and their qualitative and quantitative production of off-odours are similar to the measurements in the spoiled product.

The latter step is particularly important, as some bacteria may produce the chemical compounds associated with spoilage but are unable to do so in significant amounts, and they are thus not the specific spoilage bacteria. When stored aerobically, levels of 10^8-10^9 cfu/g of specific spoilage bacteria are required to cause spoilage. The spoilage of packed fish is seen at a much lower level of 10^7 cfu P. *phosphoreum* per gramme. This relatively low level is probably due to the very large size (5 μm) of the bacterium resulting in a much higher yield of for example, TMA per cell . Spoilage potential and activity can be assessed in several fish substrates as sterile, raw fish juice, heat-sterilized fish juice (Castell and Greenough, 1957; Gram *et al., 1987*; Dalgaard, 1993) or on sterile muscle blocks . The latter is the most complicated but is also that yielding results comparable to the product. If any of the fish juices are chosen, it is important that the growth rate of the spoilage bacteria in the model system is equal to the growth rate in the product.

A qualitative test for the ability of the bacteria to produce H_2S and/or reduce TMAO may also be used when the spoilage flora is screened for potential spoilage bacteria. A medium where the reduction of TMAO to TMA is seen as a redox indicator changes colour, and the formation of H_2S is evident from a black precipitation of FeS which has been developed for this purpose .

Shewanella putrefaciens has been identified as the specific spoilage bacteria of marine temperate- water fish stored aerobically in ice. If the product is vacuum-packed, P. phosphoreum participates in the spoilage and it becomes the specific spoilage bacteria of CO_2 packed fish. The spoilage flora on iced tropical fish from marine waters is composed almost exclusively of *Pseudomonas* spp. and *S. putrefaciens*. Some *Pseudomonas* spp. are the specific spoilers of iced stored tropical freshwater fish and are also, together with *S. putrefaciens,* spoilers of marine tropical fish stored in ice.

At ambient temperature, motile aeromonads are the specific spoilers of aerobically stored freshwater fish. Barile *et al.* (1985) showed that a large proportion of the flora on ambient-stored mackerel consisted of *S. putrefaciens,* indicating that this bacterium may also take part in the spoilage.

(1) Modified Atmosphere Packaging (CO_2 containing)
(2) LAB: Lactic Acid Bacteria
(3) Fish caught in tropical waters or freshwaters tend to have a spoilage dominated by *Pseudomonas spp.*

5

Techniques in Fish Handling and Processing

HANDLING WET FISH

Much of the fish landed in Britain is preserved by chilling in ice from the time it is caught until it reaches the consumer. Fish preserved in this way is known as wet fish. A growing proportion of the catch is frozen at sea immediately after capture, and considerable amounts of iced fish are frozen at the ports after landing; these operations are described under freezing and cold storage.

HANDLING WET FISH AT SEA

The length of voyage of fishing vessels storing the catch in ice may range from a few hours for small inshore vessels to about three weeks for the largest distant water trawlers. White fish, that is those species in which most of the fat is in the liver and the flesh is lean, are handled in much the same way on all sizes of vessel. The catch is released from the net on to the deck, gutted immediately, washed, and stowed with ice in boxes or compartments below deck.

Gutting of round fish like cod, haddock and whiting means slitting the belly from throat to vent, removing the liver and cutting out the guts to leave the belly cavity empty. This operation is traditionally done by hand with a knife, but gutting machines are coming into use on both large and small ships to make the task of the fisherman easier. Gutting helps to preserve the fish by removing the main source of spoilage bacteria and digestive juices which attack the flesh of the fish after death. On the larger fishing vessels the livers are cooked in steam boilers to extract the liver oil, but on small boats the livers are discarded with other offal.

The gutted fish are washed to remove traces of blood and debris, and to wash away most of the bacteria present on the skin and in the gills of the fish.

The washing equipment on small boats may be simply a hose and an open mesh basket, but on large trawlers a more sophisticated washing tank with circulating water is in general use. In these washers the fish are discharged over a weir and down a chute to the fishroom below deck.

Fishrooms for iced fish are mainly of two types, either an undivided hold in which the catch is stowed in boxes, or a hold divided by partitions into a number of sections called pounds in which the catch is stowed on portable shelves. The principle of stowage is the same; the fish are in shallow layers completely surrounded by ice, whether on a shelf or in a box, so that they are cooled rapidly to ice temperature and kept close to 0°C throughout the voyage. About one part of ice to three parts of fish by weight is required to protect fish for up to 5 days; one part of ice to two of fish is needed for longer voyages. White fish, promptly gutted, washed, and stowed in ample ice, will keep in first class condition for 5-6 days, become stale after 10-12 days, and are unlikely to be edible after 15-16 days.

Boxed stowage is usual on smaller fishing vessels, and the practice of boxing is gradually being extended to larger ships, since the method has a number of advantages including delivery of the fish to the merchant undisturbed by rehandling at the port market, ease of identification of size, species and time in ice, and avoidance of damage and loss of weight during stowage.

Ice plants, particularly older ones at the larger ports, supply crushed block ice to fishing vessels, but more recently built plants, particularly at smaller ports where ice was not locally available in the past, usually deliver small, smooth pieces of ice known as flake ice. Flake ice is normally bulkier than crushed block ice, but weight for weight the cooling capacity of all types of ice, made by any method and from hard or soft fresh water, is the same.

The fish are cooled when heat is absorbed by the surrounding ice, which is thus melted. Further cooling is obtained when the cold meltwater trickles down between the fish. The fishroom temperature is ideally kept slightly above 0°C in order to allow the ice to melt slowly, but is not kept so high as to waste the ice. To this end, most fishrooms on large vessels are completely insulated, and fishrooms on small boats often have partial insulation. Mechanical refrigeration plants are installed on one or two inshore vessels and on some, but by no means all, larger ones; their main purpose is to conserve ice on the outward voyage, and to keep the fishroom air cool during fishing; they have little or no direct effect on the stowed fish, which depend for cooling almost entirely on the surrounding ice.

Fatty fish, like herring, sprats, mackerel and pilchards, *i.e.* those containing a good deal of fat or oil, are not normally gutted at sea because their small size and the large numbers in which they are caught make this impracticable in the time available on typically short voyages to grounds not far from the port of landing. They are usually put below straight from the net, and iced in boxes.

The keeping time of fatty fish in ice is much less than for white fish; the attack by bacteria and digestive juices is much more rapid because the fish are ungutted, and the fat absorbs oxygen to produce rancid flavours and odours. Herring for example are normally required to be in the hands of the port processors within 1-2 days after catching to give a first class product, although for some outlets it is possible to keep herring with a low fat content for 4-5 days in ice. Stowage at sea in refrigerated sea water is a possible alternative to ice as a means of rapidly cooling large quantities of small fatty fish, although the method is not yet in general use in the UK.

FREEZING AND COLD STORAGE OF FISH

It is perfectly feasible to keep fresh fish for many months without perceptible change in the eating quality by rapidly freezing it soon after catching and then storing it at a suitably low and constant temperature. With this method of preservation, the thawed product is virtually indistinguishable from the best fresh fish. Freezing has revolutionized the fish processing industry in Britain since the Second World War.

FREEZING FISH

Heat is removed from the fish in the freezing process either by surrounding the fish with a stream of cold air, by placing the fish in contact with a cold surface or by spraying with certain liquid refrigerants. Three main types of freezing plant are used that employ these techniques, the air blast freezer, the plate freezer and the immersion freezer.

The air blast freezer is essentially a tunnel in which a fast-moving stream of very cold air is blown over the fish, which are placed on trolleys or on a moving belt. The air is usually at a temperature of-30 to -40°C and moving at about 5 m/s. The air blast freezer is most suitable for a wide range of sizes of fish, and for products of irregular shape.

The plate freezer is more compact than the air blast freezer, and is most useful for handling fish products that are uniform in thickness and that have a reasonably flat surface which can make good contact with the cold plates. Two versions of the plate freezer are in general commercial use, the horizontal and the vertical types. The horizontal plate freezer, used mainly in land installations, handles many of the catering and retail fish products that are already packed in cartons prior to freezing. Trays of packs are slid between pairs of horizontal plates, the plates are closed tightly on to the packs by hydraulic pressure to make good contact, and a cold refrigerant is circulated through serpentine passages within the plates; a retail pack 3 cm thick takes about an hour to freeze.

The vertical plate freezer, which was originally designed in the 1950s for use on fishing vessels, is employed for freezing large blocks of whole fish. The fish are packed between pairs of plates, usually without any wrappings, and the plates moved slightly towards each other to compact the block and ensure good contact. Liquid refrigerant is circulated through the serpentine passages within the plates until the fish are frozen; the complete process for a block of whole cod 10 cm thick takes about 4 hours including loading and unloading time, with refrigerant at -40°C.

The immersion freezer is not used as much as the plate and blast freezers in the British fish industry; the main difficulty is the limited choice of liquids that are both good refrigerants and suitable for use in direct contact with foods. Brines are sometimes used, but the fish may take up too much salt. Liquid nitrogen is used successfully in one type of immersion freezer, which subjects the product to a spray of liquid nitrogen as it passes through a tunnel.

The freezing process cannot improve the quality of fish; therefore the best frozen products are those made from first class raw material. This applies particularly to whole fish which, after thawing, are likely to be subjected to further processing. Whole iced cod, for example, when frozen not later than 3 days after catching can on thawing be treated in the same way as very fresh fish, but cod that has been delayed longer than this, or has been kept uniced, is unlikely after freezing and thawing to yield fillets of a high quality. Whilst some species, flatfish for example, can be kept in ice a little longer than cod before freezing without impairing the quality of the thawed product, others like haddock and hake do not keep so well. Fillets of most white fish species can be taken from whole fish 5-6 days in ice and frozen to give a high quality thawed product.

The freezing process should always be completed as rapidly as possible, not only to increase output of the equipment, but also to reduce the time in which bacteria and digestive juices are still able to attack the fish; bacterial action ceases below about -10°C, and the activity of enzymes is reduced as the temperature falls. There are marked changes in texture and flavour when fish is frozen very slowly at temperatures only a little below 0°C. The final temperature of fish being frozen should be that at which it is to be stored, namely -30°C; this ensures that the frozen product imposes no extra heat load on the cold store since this is designed only to keep the product cold and not to freeze it.

Freezing at Sea

On freezer trawlers handling whole fish, the catch is brought in over the stern, discharged to an enclosed processing deck, gutted by hand or machine, washed and bled in cold sea water and conveyed forward to the freezers, where blocks of about 50 kg are produced in vertical plate freezers and transferred to cold storage rooms running at -30°C. These ships make voyages of 4-8 weeks, carry a crew only two or three more in number than their counterparts handling iced fish, and bring back about 500 tonnes of frozen cargo for transfer to shore cold stores at the ports. Factory trawlers that fillet the catch before freezing are less numerous than trawlers which freeze whole fish. On the former the gutted fish are filleted mainly by machine, then the fillets are packed into blocks of about 10-15 kg and frozen in horizontal plate freezers before being transferred to the ship's cold store. Compared to the situation with a trawler freezing whole fish, the crew has to be larger to cope with factory operation at sea, and the quality of fillets frozen at sea as opposed to whole fish frozen at sea depends much more on careful handling and processing, but on the other hand the storage capacity of the ship in terms of frozen edible portion is considerably increased.

It is likely that the distant water fishing fleet will gradually change over completely from the stowage of wet fish in ice to the freezing of fish at sea.

Cold Storage of Fish

Temperature of storage is the most important single factor affecting the storage life of frozen fish. Almost all cold stores for fish that have been built in recent years are designed to operate at -30°C, at which temperature the products will keep in first class condition for several months. Lean fish, such as cod and haddock, when stored for long periods at too high a temperature or in fluctuating temperatures can have marked changes in texture and flavour when thawed out. The thawed flesh may feel rubbery and appear dense white instead of translucent. After cooking they are found to be tough and fibrous or stringy.

Other causes of change in cold stored fish are dehydration and oxidation. Dehydration is kept to a minimum either by glazing unwrapped frozen fish before storing them, that is covering the surface with a skin of ice by dipping them quickly in cold water, or by packaging the fish in a material that is a good barrier against the passage of water vapour, for example polyethylene film. Fatty fish like herring are particularly prone to absorb oxygen from the air and so become rancid; these are therefore wrapped in a material that forms a good oxygen barrier. In addition the space between the package and the contents may be evacuated to reduce even further the risk of rancidity. The ideal packaging material for fish products is often a laminated film combining the desired properties of two or more plastics. The cold storage chain is maintained in distribution by the use of insulated, refrigerated vehicles or containers, both of which operate at about -20°C, and frozen food cabinets, again at -20°C, in shops and catering premises.

Thawing Frozen Fish

The growth of quick freezing in the fish industry, and particularly the production of large blocks of whole fish for subsequent processing, has made necessary the development of thawing plant. There are two main types of equipment, those in which the fish are heated in a warm air stream or in warm water, and those which directly use heat generated by electricity. The method most used for large blocks of sea frozen fish is air blast thawing; the fish are conveyed through a stream of moist moving air at about 20°C until they are thawed enough to permit filleting or other processing. The fish temperature should never exceed 20°C.

DRIED AND SALTED FISH

Bacteria and moulds generally cannot grow in the absence of water and hence drying can be used as a means of preservation. Salt, if present in sufficient strength, will slow down or prevent bacterial spoilage of fish. Drying or salting, or a combination of both, have been used in the fish industry for centuries but nowadays only a very small proportion of the catch in Britain is processed by these methods.A few companies still make dried salted fish from cod and related species, by heading and splitting the fish, removing most of the backbone, and stacking the fish in piles with layers of salt between them. The juices withdrawn from the fish by the salt are allowed to run away and, after frequent restocking over a period of months, the water content is further reduced by drying the fish in a heated chamber until the moisture content is somewhere between 10 and 30 per cent.

Some herring are pickle cured, particularly in north-east Scotland and the Shetlands, although this export trade is a mere shadow of what it was earlier this century. The whole herring, having been lightly sprinkled with salt while awaiting processing, are first gibbed by hand or machine, that is the gills, long gut and stomach are removed. They are then packed in barrels with a layer of salt on each layer of herring until the barrels are full. After a day or two, when the herring have shrunk appreciably, the barrels are topped up with further layers of herring and salt, the lids are fitted and the barrels then laid on their sides for 8-10 days. After this period the barrels are up-ended, the lids removed, and the blood pickle from the upper half drained off through the bung-hole. To make ready for storage the barrels are finally topped up with fish, the lids replaced and blood pickle poured through the bung-holes until all spaces are filled. Klondyking is the name given to another method of preserving ungutted herring using salt. The name is thought to originate from a method developed about the same time as the famous Gold Rush of 1897.

The herring, contained in baskets, are sprinkled with coarse salt, approximately 175 kg of fish treated with about 10 kg of salt, as they are tipped into a wooden box. Ice is then put on this mixture offish and salt, about 100-125 kg of ice to every 175 kg of mixture. Herring treated in this way could be kept in edible condition for approximately 1 week.

CANNED FISH

The canned fish industry in Britain is a small one, and the range of products is confined mainly to herring, sprats and pilchards packed in either tomato sauce or vegetable oil. The process for herring in tomato sauce is typical. First the fish are nobbed by machine, that is the head and gut are removed. They are then immersed in saturated brine for up to 30 minutes and packed by hand into oval cans holding 200 g of fish. The tomato sauce is added, the can lids are lightly clipped on and the cans exhausted in steam for 10-15 minutes. The object of exhausting is to produce a partial vacuum in the headspace of the can which is not filled either with solids or liquids. The cans are then sealed, washed, and heat processed in steam at 115ºC for 55 minutes. After cooling, the cans are stored for about a month, labelled and packed in outer cartons for dispatch. The heat processing stage is critical, and is designed to inactivate all bacteria and enzymes present and in particular to destroy any harmful organisms.

There are a number of imported canned fish products that could be manufactured in this country from British-caught fish, and the canning industry is examining the possibilities of expanding their range.

HANDLING WET FISH ON SHORE

The catches of British fishing vessels are normally auctioned to fish processing firms at the fishing ports. Most of the large ports are on the east coast, including the three biggest. Hull, Grimsby and Aberdeen. Covered quayside markets are provided at almost all the ports, and merchants' premises are usually adjacent to, or a short distance from, the markets. At a few remote landing places, particularly on the west coast of Scotland, where no local processing facilities are available, the catches are discharged and consigned by road to the processing centres on the east coast.

Comparatively small amounts of white fish are dispatched unprocessed, repacked in fresh ice, to inland destinations but most white fish is filleted in premises at, the port. Although machines are available for filleting most species, a large proportion of the catch, particularly when handled by the smaller processing firms, is still filleted by hand. The fillets, which constitute roughly 40-50 per cent by weight of the gutted whole fish, are either packed in ice in non-returnable boxes and sent to inland wholesalers and retailers, or are further processed at the port, mainly by smoking or freezing, or both.

A high proportion of the herring catch goes to the kipper trade, and herring for this outlet is split or filleted by machine for making kippers and kipper fillets respectively.

The traditional container for inland carriage of fillets has been the non-returnable wooden box, mainly in units of 7 and 14 lb, but this has been superseded in many large firms by a waterproof fibreboard box and, to a lesser extent, by an expanded polystyrene box, which has some advantage as an insulated container but is more susceptible to damage due to rough handling during transit. Fillets should be kept close to 0°C during transit, and in good practice the fillets are packed in thin layers with ice top and bottom and a little more ice sprinkled among the fish.

Fish is nowadays mostly carried by road transport from the ports and the biggest companies have their own fleets of insulated and refrigerated vehicles carrying fish either to inland distribution depots or direct to customers, while many smaller merchants at the ports share a long-distance transport pool.

Although the better carriers use vehicles with adequate insulation, often supplemented by mechanical cooling units, some fish is still carried under less satisfactory conditions on open lorries; much greater reliance has then to be placed on ice and on the insulating properties of the boxes to protect the contents during distribution. Under the best conditions, wet fish can be on sale to the consumer anywhere in Britain within twenty four hours of landing; a typical timetable would be discharge from the fishing vessel in the early hours of the morning, sale to the port merchant and a short journey to his premises at about 8 am, filleting, packing and icing by midday and dispatch by road from the port in the afternoon to arrive at an inland depot in the early hours of the next day, from where it is delivered to the retail shops in time for that day's trading.

Most retailers keep the bulk of their supplies in chillrooms on the premises and only display a selection on ice, often in refrigerated cabinets.

The following Advisory Notes expand the information on handling wet fish on shore: 1 *The care of the fishmongers fish,* by G. H. O. Burgess; 3 *The handling of wet fish during distribution;,* 10 *Fishworking premises- materials and design,* by J. J. Waterman; 12 *Fish display in retail shops;* 16 *Non-returnable fish boxes,* by J. Wignall; 17 *Measures, stowage rates and yields of fishery products,* by J. J. Waterman; 23 *Control of flies in fishmongers' shops;* 42 *Fish for caterers and friers,* by J. C. Early and R. Malton; 45 *Cleaning in the fish industry,* by I. N. Tatterson and M. L. Windsor.

Smoked Fish

Fish is smoked nowadays mainly to give it a pleasant flavour rather than to preserve it. Present day products are therefore only lightly salted and smoked and will not remain edible for much more than a week at ordinary temperatures. The smoking process consists of passing wood smoke over the surface of the fish, in a kiln. Most British products are cold smoked, that is the fish remains uncooked and the kiln temperature does not rise above 30°C. Typical smoked products are the finnan haddock, smoked cod fillet, the golden cutlet and the kipper. Fish that are hot smoked are cooked during the process; the kiln temperature may be as high as 80°C and the fish temperature may reach 60°C. Some hot smoked products in this country are sprats, eels, trout, buckling made from herring, and Arbroath smokies made from small haddocks.

Two types of smoking kiln are in general use, the traditional chimney kiln and the Torry mechanical kiln. It is estimated that more than half of the smoked fish made in Britain is now produced in mechanical kilns, and the proportion is continually increasing.

Before smoking, the fish are immersed in a brine solution. This assists in removing some of the water in the fish, thus tending to firm the flesh. The salt imparts a flavour to the product, but concentration and purity of the salt are extremely important and require to be carefully controlled. A 70 to 80 per cent saturated brine is used in most modern smoke cures.

Following the salting treatment, pre-drying of the fish is required in order to remove some of the moisture prior to smoking. For this purpose, the fish are hung to drip on open racks.

The source of smoke is almost universally a smouldering fire of hardwood chips and sawdust; although more sophisticated smoke producers have been made from time to time, and are used for smoking other foods, these have so far made little impact on the fish trade. In the traditional chimney kiln, the open fires are at the base of a tall, brick-built structure in which the fish are hung on rails of various types called banjoes, speats or tenters, and thus exposed to the rising smoke and warm air. The repositioning of the fish during smoking and eventual removal of the finished products are slow hand operations which require the services of a skilled craftsman in order to produce a satisfactory article.

In the mechanical kiln, the fires are contained in separate fireboxes, and the smoke is blown horizontally through trolleys holding fish in the kiln; the fish may be hung on rails or laid on trays, either of which are supported in the trolley. The temperature and speed of the mixture of smoke and air is carefully controlled to give a uniform product throughout the kiln in a much shorter time than is possible in the chimney kiln. Fish handling is much reduced using the mechanical kiln which can also readily be incorporated in the factory production line. Partial drying as well as smoke deposition is an essential part of the smoke curing process; typically a kipper which should lose about 14 per cent in weight during smoking will require 6-12 hours in a traditional kiln, but only 4 hours in a mechanical one.

Most cold smoked fish products are only lightly coloured by the smoke, and so a permitted dye is normally added to the brine bath through which the fish pass before going into the kiln, in order to enhance the appearance of the finished product.

HANDLING SHELLFISH

The shellfish industry, although only a small part of the fish industry as a whole, has grown considerably in recent years, and the products are generally high value ones. The principal species in order of importance are Norway lobster or scampi, lobster, scallop, crab, cockle, crawfish and oyster. Mussels and shrimp make only small contributions to British landings at the present time, but their fisheries are capable of considerable expansion. The Norway lobster is landed either whole or headless in ice, and is often frozen in the shell while awaiting processing. The meats are extracted from the thawed tails either by hand peeling or by blowing out with air or water jets. The peeled meats are frozen individually, either in an air blast freezer or a liquid nitrogen freezer, glazed and bagged and put into cold storage.

Lobsters are still distributed live inland; the few that are processed are normally cooked and then frozen whole. Crabs, which do not travel well, are processed close to the points of landing. They are boiled whole, and the meats are then extracted from body and claws by hand. The white and brown meats are frozen separately and then wrapped and cold stored. A small amount of crab meat is also canned in Britain, and small amounts of cooked whole crabs are distributed chilled to retailers.

The oyster is marketed live in shell, and the crawfish catch is mainly exported live to the Continent. The scallop, and its near relative the queen, are frozen in shell on arrival at the port processing plant, thawed as required and the meats removed by hand with a knife. The meats are frozen individually, usually in an air blast freezer, and bagged for cold storage; they are mainly exported to the United States. Cockles and mussels are boiled in the shell at the ports; the extracted meats may be distributed either chilled, frozen, or packed in jars in brine or vinegar.

Shrimps are cooked and peeled soon after catching, and the meats distributed either chilled, frozen or potted in butter; alternatively the whole raw shrimp are iced and then frozen for subsequent processing.

FISH BYPRODUCTS

When the edible flesh is removed by filleting from whole fish landed for human consumption, the remaining heads, skeletons and other processing waste are disposed of by making fish meal. In addition whole fish, both white and fatty, that are surplus to market requirements or landed specifically for the byproducts industry are also converted into fish meal and oil. White fish and white fish offal has most of the water removed by cooking and drying, and the dried material ground and bagged for use as a high protein animal food, particularly in the pig and poultry industries. The yield is roughly one tonne of meal from 5 tonnes of raw material. Fatty raw material is pressed after the cooking stage and the press liquor, a mixture of oil, water and some solids, is further processed to separate the oils and solids. The oil-free solids are dried and ground along with the bulk of the solid material from the press. The refined fish oils make a valuable contribution to the manufacture of edible products like margarine.

Fish meal made from suitable raw material under hygienic conditions can be eaten by humans, and its possible use as a fish concentrate protein added to the diet of protein-deficient peoples in the developing countries is being investigated in many parts of the world. Waste from the British shellfish industry is occasionally accepted in small quantities by the fish meal manufacturers, but most of the shells are at present unsuitable for the manufacture of byproducts.

Only a small part of this country's requirements for fish meal are met by the British industry; the possibility of reducing imports by increasing the British catch of species unwanted for human consumption and converting it into fish meal is at present being explored.

THE IMPORTANCE OF CONTAINERS IN FISH HANDLING

Containers of various types, sizes and of different materials are used all over the world to hold fish both on board vessels, in processing, during transport and under general storage. Poor handling and lack of suitable containers leads to as much as 20-30% spoilage in many countries.

Containers are used to perform the following functions:

1. ease the handling of small and large quantities of fish;
2. simplify and increase the speed of unloading/loading and transportation of raw material;
3. protect the fish against physical damage contamination and other deteriorating factors;
4. offer a suitable packing unit for fish and ice;
5. contain the fish under such conditions that it reaches the buyer in the best possible condition;
6. help to protect the raw material against natural deteriorating effects;
7. help to make maximum utilization of resources and to achieve optimum economical results through the whole system of handling from harvest to consumption.

In the whole system of correct fish handling the container is only one factor in the process. Knowledge of handling, necessary regulations/laws and the specific behaviour of fish meat under various conditions are also important points to consider.

Various packing materials and containers have their limitations. It is therefore important to choose the right container system at the right time and to use it in accordance with laid down objectives.

In conclusion there is a great variation in the size, design and construction of containers used in various parts of the world. The materials used depend on the type of fishing, size of vessel, type of road transport, degree of organization of the industry, value of the catch, and sometimes local traditions.

DEFINITION

In this report it is felt necessary to define a difference between fish baskets, boxes and containers.

Fish Baskets

Originally a container of plaited or woven material. Most often made of cane or some other strong plant fibre. This type is still commonly in use in most developing countries. In the industrialized fishing nations the traditional plant fibre has been replaced by plastic. It has been common to make the fish basket in sizes ranging from units which could take around 10 kg up to those able to take 100 kg.

The basket has gradually been losing ground to the fish box due to its unsuitable shape for storage (in fish holds, cold stores, etc.) and for mechanized handling, by fork lifts and similar systems.

Fish Boxes

Generally these are uninstalled containers with a capacity from 10 litres to around 100 litres. Each unit is designed to enable handling by 1 or 2 persons.

The fish box has its application on board fishing vessels, in warehousing, processing, transport and at the fish market.

The 3 main areas for use of boxes are as follows:

a. Tote boxes used aboard fishing vessels and for transport of fresh fish and storage of raw material awaiting processing. Boxes for these uses have to be strong, stable, easy to stow, simple to clean and preferably of the returnable type.
b. Boxes for internal use in the processing industry. Normally such boxes are of a lighter and smaller type. Since these units will be containing semi-processed or processed products the hygienic requirements will be high. Most of them are of the returnable type and made of aluminium or plastic. Size, 10 to 30 litres.
c. Boxes or packing material for marketing. At this stage non-returnable boxes are generally used. The main requirement is that the packing material should have sufficient strength to protect the fish until it reaches the consumer, but be low priced. Most often boxes for this use are made of styrofoam, plastic, light wood, or fibre board.

Fish Containers

These are insulated or uninstalled units often tailored to a special purpose or use. Most often they have their application in storage on board vessels and ashore and in transport of fish.

a. Containers used for cooling with ice. Preferably such types should be insulated and have an insulated lid. Size may vary from 50 up to 1000 litres. In this type of container fish can be held for a considerable period of time if sufficient ice is used. They have their application on board fishing vessels, in transport of fish (sea, land, air), as storage units for the processing industry and in the fish market.

b. Self contained cooling containers. Over the past 20 years containers used for transport of various products have been standardized. For transporting products that must be held at low temperatures, insulated, refrigerated containers have been developed. These containers have been designed to standards set by the International Standarization Organization (ISO) so as to fit into the general goods transport system developed for sea, road, railway and air transport.

Containers used for transport of fish can take from 5 to 30 t and are mostly used for long distance transport.

The general requirements for all packaging material used for packing of food, fish included, are that the packaging material shall be clean, undamaged and suitable for the purpose.

The physical strength of the material must be so good that the commodity is protected against pressure and shocks in the period the commodity is stored.

Further from a chemical point of view the materials must fulfill the following requirements:

- must not contain or produce poisonous or health damaging substances or compounds that can be transferred to people or animals eating the fish;
- must not produce smell, taste or colour which can contaminate the product.

To fulfill these requirements it is important that a set of official rules be formulated and that these rules are enforced by the appropriate authorities.

BASICS OF FRESH FISH HANDLING AND USE OF ICE

Throughout history, man has preferred to consume fresh fish rather than other types of fish products. However, fish spoil very quickly and man has had to develop methods to preserve fish very early in history.

Keeping and Transporting live Fish

The first obvious way of avoiding spoilage and loss of quality is to keep caught fish alive until consumption. Handling of live fish for trade and consumption has been practised in China with carp probably for more than three thousand years. Today, keeping fish alive for consumption is a common fish-handling practice both in developed and developing countries and at both artisanal and industrial level.

In the case of live fish handling, fish are first conditioned in a container with clean water, while the damaged, sick and dead fish are removed. Fish are put to starve and, if possible, water temperature is reduced in order to reduce metabolic rates and make fish less active. Low metabolic rates decrease the fouling of water with ammonia, nitrite and carbon dioxide that are toxic to fish and impair their ability to extract oxygen from water. Such toxic substances will tend to increase mortality rates. Less active fish allow for an increase in the packing density of fish in the container. A large number of fish species are usually kept alive in holding basins, floating cages, wells and fish yards. Holding basins, normally associated with fish culture companies, can be equipped with oxygen control, water filtering and circulation and temperature control. However, more simple methods are also used in practice, for instance large palm woven baskets acting as floating cages in rivers (China), or simple fish yards constructed in a backwater of a river or rivulet for large "surubi" (Platystoma spp.), "pacu" (Colossoma spp.) and "pirarucu" (Arapalma gigas) in the Amazonian and Parana basins in South America.

Methods of transporting live fish range from very sophisticated systems installed on trucks that regulate temperature, filter and recycle water and add oxygen, to very simple artisanal systems of transporting fish in plastic bags with an oxygen supersaturated atmosphere . There are trucks that can transport up to 50 t of live salmon; however, there is also the possibility of transporting a few kilo-grammes of live fish relatively easily in a plastic bag.

By now a large number of species, inter alia, salmon, trout, carp, eel, seabream, flounder, turbot, catfish, Clarias, tilapias, mussels, oysters, cockles, shrimp, crab and lobster are kept alive and transported, very often from one country to another.

There are wide differences in the behaviour and resistance of the various species. Therefore the method of keeping and transporting live fish should be tailored according to the particular species and the length of time it needs to be kept outside its natural habitat before slaughtering. For instance, the lungfish (Protopterus spp.) can be transported and kept alive out of water for long periods, merely by keeping its skin moist.

Some species of fish, noticeably freshwater fish, are more resistant than others to changes in oxygen in solution and the presence of toxic substances. This is probably due to the fact that their biology is adapted to the wide yearly variations in water composition presented by some rivers (cycles of matter in suspension and dissolved oxygen). In these cases, live fish are kept and transported just by changing the water from time to time in the transport containers. This method is widely used in the Amazonian, Parana and Orinoco basins in South America; in Asia (particularly in the People's Republic of China, where also more sophisticated methods are used) and in Africa .

Aluminium containers with live freshwater fish are stored in the aisles of a public transport vessel. Containers are covered with palm leaves and water hyacinth to prevent the fish from jumping out of the containers and to reduce evaporation. The water in the containers is changed from time to time and an almost continuous visual control is kept on fish. Dead fish are immediately put to smoke-drying (African style) in drum smokers, also transported in the vessels or transporting barges.

The most recent development is the keeping and transporting of fish in a state of hibernation. In this method, the body temperature of live fish is reduced drastically in order to reduce fish metabolism and to eliminate fish movement completely. The method greatly reduces death rates and increases package density, but careful temperature control should be exercised to maintain the hibernation temperature. There is an appropriate hibernation temperature for each species. Although the method is already utilized for instance to transport live "kuruma" shrimp *(Penaeus japonicus)* and lobster in pre-chilled wet sawdust, it should be considered an experimental technique for most of the species.

Although keeping and transporting live fish is becoming more and more important, it is not a viable solution for most of the bulk fish captures in the world.

Chilling Fish with Ice

Historical evidence proves that the Ancient Chinese utilized natural ice to preserve fish more than three thousand years ago. Natural ice mixed with seaweed was also used by the Ancient Romans to keep fish fresh. However, it was the development of mechanical refrigeration which made ice readily available for use in fish preservation. In developed countries, particularly in USA and some European countries, the tradition of chilling fish with ice dates back more than a century. The practical advantages of utilizing ice in fresh fish handling are therefore well established. However, it is worthwhile for young generations of fish technologists and newcomers to the field, to review them, paying attention to the main points of this technique.

Ice is utilized in fish preservation for one or more of the following reasons:

Temperature Reduction: By reducing temperature to about 0°C the growth of spoilage and pathogenic micro-organisms is reduced, thus reducing the spoilage rate and reducing or eliminating some safety risks.

Temperature reduction also reduces the rate of enzymatic reactions, in particular those linked to early *post mortem* changes extending, if properly applied, the *rigor mortis* period.

Fish temperature reduction is by far the most important effect of ice utilization. Therefore, the quicker the ice chills the better. Although cold-shock reactions have been reported in a few tropical species when iced, leading to a loss of yield of fillets, the advantage of quick chilling usually outweighs other considerations. The development of ad hoc fish handling methods is of course not ruled out in the case of species that could present cold-shock behaviour.

Melting Ice keeps Fish Moist: This action mainly prevents surface dehydration and reduces weight losses. Melting water also increases the heat transport between fish and ice surfaces (water conducts heat better than air): the quickest practical chilling rate is obtained in a slurry of water and ice (*e.g.*, the CSW system).

If, for some reason, ice is not utilized immediately after catching the fish, it is worthwhile keeping the fish moist. Evaporative cooling usually reduces the surface temperature of fish below the optimum growth temperature of common spoilage and pathogenic bacteria; although it does not prevent spoiling. Ice should also be utilized in relation with chilling rooms to keep fish moist. It is advisable to keep chilling room temperature slightly above 0°C (*e.g.*, 3-4°C).

However, water has a leaching effect and may drain away colour pigments from fish skin and gills. Ice melting water can also leach micronutrients in the case of fillets and extract relatively large amounts of soluble substances in some species (*e.g.*, squid).

Depending on the species, severity of leaching and market requirements, an ad *hoc* handling procedure may be justified. In general, it has been found that drainage of ice meltwater is advisable in boxes and containers and that permanence of fish in chilled sea water (CSW) and refrigerated seawater (RSW) should be carefully assessed if leaching and other effects (*e.g.*, uptake of salt from the seawater, whitening of fish eyes and gills) are to be avoided.

During the past there was much discussion about allowing drainage from one fish box to another, and consequent reduction or increase of bacterial load by washing with drainage water. Today, apart from the fact that in many cases box design allows for external drainage of each box in a stack, it is recognized that these aspects have less importance when compared with the need for quick reduction in temperature.

MATERIALS TO BE USED IN FISH BOXES AND CONTAINERS - PROPERTIES AND SUITABILITY

In manufacturing of fish boxes and containers, a great number of different materials have been used. The main categories of materials are metals or alloys, natural materials like wood, bamboo, fibre board, etc., and synthetic materials (plastics).

The most common materials used in manufacturing fish boxes and insulated fish containers are wood, aluminium alloys, HD-polyethylene (high-density polyethylene), expanded polystyrene ("styropor") and solid fibre board. Even if other materials such as steel, bamboo, etc., are in common use in many developing countries, the properties of these materials are not very suitable. Steel is very corrosive and is three times denser than aluminium. Bamboo, as it is normally used, is a material which presents hygiene problems. Materials are rated on their cost, ease of constructing suitable containers and physical and chemical proportion.

PROPERTIES OF MATERIALS USED FOR FISH BOXES AND INSULATED CONTAINERS

Plastic

Fish boxes and containers made from HD-polyethylene are superior to those made from other materials, but they are rather expensive. The fabrication is inexpensive, but the material carries a relatively high cost (depending on oil prices). Production must be carried out in large numbers to get a price which competes with other materials. The expected life is 5-7 years when properly handled. Life is longer for boxes designed to facilitate stocking.

The material is impervious to liquids which facilitates cleaning and reduces opportunities for bacterial growth. However, the material is easily changed with static electricity which can draw dust when stored for long time. The temperature limitations are +100 °C and -40°C, but it must be realized that the material becomes brittle at low temperatures. Therefore it is not suitable as packing material for freezing.

Evaluation of physical and chemical properties of HD-polyethylene:

- Weight: The density of HD-polyethylene i8 low compared with many other materials.
- Corrosion: The material is resistant to rust and corrosion.
- Insulation effect: HD-polyethylene 15 a poor conductor of heat and provides a better insulation than metals.
- Workability- fabrication: HD-polyethylene is adapted to mass production methods and can be produced in large quantities at low manufacturing costs. On the other hand complexity of production methods reduces the opportunity to produce fish boxes without professional skills and equipment.
- Noise: The material makes some noise when it is handled but is less noisy than some materials such as metals.
- Toxity and resistance to chemicals: The material is not toxic and transfers no smell or taste to the contents. It is resistant to oil and grease and cleaning chemical.
- Surface: The surface is smooth and easy to clean.
- Colours: Products in a wide range of colours are available.
- Strength: HD-polyethylene has only medium strength and has disadvantages with respect to mechanical damage.
- Repairability: Broken boxes and containers are generally not repairable and must usually be replaced.

Aluminium Alloys

Fish boxes and containers made from aluminium alloys are expensive. They are manufactured in forms suitable both for stacking and nesting. and are well suited for freezing of fish.

Physical and chemical properties to be considered:

- Weight: Aluminium is known for its lightness compared with other metals (approx. one third of the weight of steel).
- Corrosion: Generally the corrosion resistance of aluminium is very good. Some foods react with aluminium. especially shellfish. (A protective lining/coating is recommended to prevent contact in such cases).

- Insulation effect: Aluminium has higher heat conductivity than other materials used for manufacturing fish boxes and containers. The insulation effect is therefore very poor. Faster freezing. however. is an advantage of aluminium.
- Workability-fabrication: Aluminium has good workability and is easily formed. Mass production is preferable.
- Strength: Aluminium boxes have a high mechanical strength.
- Toxity: Aluminium is non-toxic and is recommendable in direct contact with most foods but can reach with some shellfish.
- Surface: The surface is smooth. and is easily cleaned. although care must be taken with alkaline detergents and sanitizers.
- Repairability: Fish boxes and containers made from aluminium alloys are not easily mended. but in certain cases it is possible to repair them.
- Durability: Expected life is 8-10 years.
- Scrap value: When no longer fit for use. the aluminium fish boxes and containers have some scrap value.
- Noise: Fish boxes and containers made from aluminium are very noisy when handled empty.

Wood

Compared with modern materials as aluminium alloys and HD-polyethylene. wooden fish boxes and containers are rather cheap. -and are used mostly in less industria1ized countries.

Wooden boxes can be used on vessels. in chill stores. in processing plants and in transport on trains and lorries. The capacity ranges from 20 to 150 kg.

Chemical and physical properties:

- Permeability: unsealed wood can absorb 15-20% water.
- Workability-fabrication: Fish boxes and containers made from wood can usually be produced near the plant in any quantity required.
- Repairability: Easy to repair.
- Toxity: Wood is a non-toxic material.
- Insulation effects: Wood gives a limited insulating effect which is lessened when wet.
- Surface: Wood has a surface that is difficult to clean thoroughly. Dealing with paint helps the cleaning. Unpainted wooden boxes are difficult to clean because of porous and rough surfaces.
- Durability: Wood has a short life-time (up to 1-2 years).
- Weight: The weight of the fish box will differ with the amount of water absorbed.

Disadvantages of Wooden Boxes

Difficult to clean. if handled roughly, they are easily damaged (mostly caused by inherent defects in the wood, poor design and poor nailing).

Expanded Polystyrene

Chemical and physical properties:

- Strength: The mechanical strength depends on the fusion process when made, but the material has less strength than other materials previously mentioned.
- Repairability: If the box is broken it cannot be repaired.
- Insulation effects: The material has superior insulation properties.
- Weight: The material is very light (down to 1/4 of wood).

Fibre Board

Boxes made of massive fibre board are primarily used as non returnable units. The fibre board is often coated with polyethylene inside and outside. These boxes are weak compared to wood, plastic and aluminium if subject to pressure.

List of properties to be considered:

- Strength: Boxes made from fibre board can sustain most types of handling except stacking without special arrangement.
- Insulation effects: The insulation properties of fibre board are comparable with plastic.
- Weight: The material is rather light (between wood and expanded polystyrene).

HYGIENE REQUIREMENTS FOR DIFFERENT PACKING MATERIALS

Boxes, containers and other equipment which come in contact with fish during catch, transport and production, have to be cleaned as often as possible and at least in between each use.

Besides the natural micro flora found on fish and shellfish, external microorganisms from the surroundings are added. Bacteria and fungi come through the soil. the air and the water as well as from human contact with raw material. Slime from fish provides a growth medium for most microorganisms. When the fish is dead, all defence mechanisms stop, and as a result the microorganisms living in the slime start to grow and attack the skin and the flesh. Offensive odours and flavours, discoloration and poor textures is the final result.

The rate of spoilage can be reduced by quick cooling and careful handling of the raw material and use of clean equipment which minimize bacterial contamination.

WATER AND WATER QUALITY

Water is the cheapest and most convenient cleaning agent to use, but dried organic soil is difficult to remove by simply scrubbing with water, so detergents are added to dislodge, disperse and dissolve soil. The quality of water used in the fishing industry is an important consideration because both chemical and bacterial contamination of seafoods can occur if polluted water is used. The level of dissolved minerals can also affect the solubility of cleaning and sanitizing agents in water.

Hard water contains elevated levels of calcium and magnesium ions. This water usually comes from limestone areas or from deep wells. The ions in hard water complex out detergents and so either a water softener must be added to remove the ions or excessive amounts of detergents will be needed for effective cleaning. Rainwater and water from lakes and streams is usually softer, containing less mineral salts Contamination of lakes and rivers from various pollution sources can render it unsatisfactory for use. In fishing, this water is used not only for cleaning but also for ice manufacture and direct human consumption, so its quality must be known.

Requirements to Water Quality

Bacteriological Requirements

- Total acceptable plate count, max. 50 bacteria per ml;
- if there are between 2-23 coliform bacteria per 100 ml, the water and water source must be given a detailed examination before being used;
- water with more than 23 coliforms per 100 ml is unsuitable for use.

Physical Requirements

- Colour = without colour (max. 20 mg Pt/l)
- Smell and taste = none, or very little
- appearance = clear and without sediments (max. 10 mg SiO_2/l)

Chemical Requirements

- Permanganate-numbers (indicates content of organic substances) max. 20 mg kMnO4/l
- pH (acidity) 6-8.0
- conductance, max. 300 microsiemen
- sulphide, max. 0.1 mg/l
- Manganese, max. 0.05 mg/l
- copper. max. 0.05 mg/l
- hardness, preferably below 2° dH

These are standard values for a good water quality used for cleaning. Chlorination, filtration, and softening agents can improve water quality.

Detergents

A great number of detergents are found in the market. These substances are often complex and contain many different chemical components. Most of the industrial detergents are corrosive and can, when used in a wrong way, cause damage to equipment as well as to persons. The buyer of detergents has to make sure that the producer gives a thorough description of the chemical composition, areas of use and the correct dosing. Information about storing, safety requirements and marking as well as information on measures to be taken in case of accidents should be included.

Requirements to Detergents

The cleaning solution must remove all dirt from the item being cleaned, without damaging it, even after it has been cleaned several times.

The detergent must be stable for storage, *i.e.*, not change appearance form or quality when stored and be quickly and completely dissolvable.

Liquid detergents are easy to use, but expensive. Powder detergents are cheaper, but require more time for accurate dosing and dissolving.

The Cleaning Process

The cleaning process can be divided into two:

1. Washing
2. Desinfection

It is not possible to use these two processes in combination and get a satisfactory result. This is clearly illustrated in the following example, from tests in the food technology industry:

Table Effect of cleaning systems on bacteria

	Bacteria numbers/cm²
Before cleaning	2 900 000
After cleaning without detergent	40 000
After washing with detergent	5 500
After desinfection	10
Desinfection without cleaning	1 200 000

A well prepared washing programme has to include the following processes:

1. Mechanical removal of dirt (brush or other equipment)
2. Rinsing with tempered (or cold) water
3. Washing with hot water (40-70°C) with detergent
4. Rinsing with tempered/cold water
5. Desinfection of the clean surface leaning time for the sanitizer to work

Even if a cleaning programme can be set up in general terms, the variations will be many because of types of contamination to be removed, type of equipment to be cleaned and available cleaning equipment.

Cleaning Methods

Because dried soil is very difficult to remove, hasing and mechanical scrubbling should be performed as soon as possible before the soil dries. This will reduce the effort needed later on. There are many methods which can be used, and a satisfactory result can be obtained in several ways. According to the size of the equipment to be cleaned, the location and the frequency of cleaning needed, the following methods can be used:

1. Manual cleaning
2. High pressure washing
3. Foam wash

Manual Cleaning

Often, the working area is small or overloaded with other equipment so that manual cleaning is the only applicable method.

Brushes of various types are simple to use and will give a satisfactory result in most cases. It is important that the equipment is kept clean and not used for other work.

This simple equipment is low priced and well suited for small operations or when the labour cost is low.

If dirt has been allowed to dry, the equipment to be cleaned should be put in water to soak for some time. Washing by hand will then be more simple and will in most cases give a good result.

High Pressure Washing

Equipment for this process is found in different shapes and sizes of both portable and stationary types. The small portable units are often adapted to small production units and are self contained.

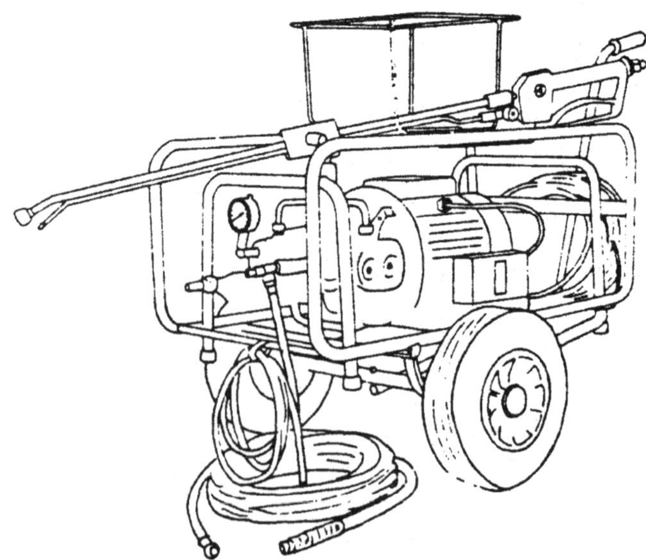

Fig. High pressure washer

Foam Washing

This process is often used in the food industry. Special foaming detergents are sprayed onto the surface to be cleaned and left in contact with that surface to allow the detergent to dissolve the soil. This is particularly useful on processing equipment where the complexity of design makes manual cleaning impossible. The foam and soil is removed by hasing or manual cleaning. Unfortunately the "airy" consistency of the foam means that it soon assumes the temperature of its surroundings so the extra cleaning power of hot detergent is not possible using this method.

To make the foam, special equipment is needed. This is connected to a tank with special chemicals and the foam/water mixture is sprayed on the items to be cleaned.

Total weight:	67 kg
Working pressure:	20.130 bar
Water capacity:	13 l/min
Water connection:	3/4 in hose
E1. Motor:	3 kW 220/380 V-3 phases
E1. cable:	8 m
H.P. hose:	8 m
Price:	approx. US$ 1 000

The stationary ones will be connected to pipe lines leading to strategic places. These units usually supply hot detergent water at pressure. The small portable units give up to 40-110 litre liquid per minute and work under a pressure of 40 kg/m². Some of these units have mechanical equipment for dosing and mixture of detergents and water. The large stationary pumps can give up to 300-litre liquid per minute under a pressure of approx. 55 kg/cm². When using such equipment the detergent can be dosed/mixed manually or automatically.

Desinfection

Washing and desinfection must be performed separately for maximum effect. There are two reasons for this. Detergents and sanitizers are often chemically incompatible and most sanitizers are complexed not only by microorganisms but also by any organic soil present.

The washing process should remove all the soil and some of the microorganisms. This should be followed by thorough rinsing to remove all detergent then application of a sanitizer or disinfectant which will kill most microorganisms and inhibit the growth of others. This should be left some time to allow it to work before rinsing off.

Desinfection is not only done to boxes and containers, but to all other equipment as well as floors and walls.

Below follows a rough list of existing types of desinfection products.

1. Chloride products
   a. hypochlorite (10-20 ppm/litre of water)
   b. chlorine amines
   c. clorisocyanur acids
2. Jodine products
3. Phenoles
4. Alcohols
5. Quaternary ammonium compounds

A variety of methods are available for application of sanitizers; these include manual spreading, spraying and fogging for enclosed tanks containers, etc..

Cleaned boxes made of modern materials like plastic, ought to have no more than 10.20 nos of bacteria per cm^2 on the surface when being examined.

FISH BOXES IN DEVELOPING COUNTRIES

In general in developing countries there is a confusing situation with a great variety of boxes and baskets being used in the fishing industry. This mixture of traditional local containers with modern plastic and aluminium boxes presents great difficulty for handling systems.

Use of Boxes/Containers

Boxes and containers of various types and materials are found in developing countries.

The degree of use and standardization are decided by the importance of the fishing industry in the national economy and the direct contact with industrialized fishing nations.

In general containers used fall into the following groups:

- Boxes (plastic, aluminium, wooden, etc.)
- Baskets (plastic)
- Various (containers/tubs/buckets/trays of steel and wood and cut off oil drums. Different surface treatments are used *e.g.* galvanizing, paint and untreated),
- Insulated containers for special use (containers of various materials without/with insulation designed for use on vessels, processing units, trucks and fish markets/shops. Most of these are custom made, but standard types are available)

Dimensions

Sizes vary a great deal. but some plastic fish boxes follow the international standards, while in other cases any available box (bread, milk, etc.), will be used. Most uninsulated containers will take from 5 up to 70-80 kg, while the insulated ones range in size from a capacity of a few kg up to 20-30 t.

Traditional Boxes and Containers

Usually the availability of raw materials, quantities of fish landed and level of fishing operation are the factors which determine the types of containers which prevail in various regions/countries. The following types of traditional containers are most common. but great variations occur:

- Leaf baskets (banana, palm tree, etc.)
- Fibre baskets (bamboo, ratten, etc.)
- Wooden boxes

Dimensions

These types of boxes/baskets come in different sizes varying from one area to another, but in general they will hold anything from around 1 kg of fish up to 60-70 kg. No standard measurements can be given.

Fig. Two Types of Traditional Fish Baskets used for Fish (Africa and Asia)

Suitability and Availability of Local Materials or Boxes and Baskets

Availability of Local Materials

The availability of raw materials for traditional baskets/boxes is in most cases sufficient as long as the demand does not increase dramatically. In some places the supply of the right type of wooden boards for boxes might cause problems, but this is more a planning and ordering problem than real shortage.

Suitability of Local Materials

Most local materials with the exception of wood can only be regarded as suitable for non-returnable or short lived returnable units.

The way these materials have been used and the traditional handling of boxes/baskets have created a reputation of the material not being hygienic. Also most traditional materials with the exception of wood are weaker and less durable than plastic and metals.

Since traditional materials most often are available within easy reach of box/container producers and the production quite often contributes significantly to the local industry, it is important that the use of traditional containers is not disregarded completely. Instead, with a more thorough approach to design and possible improvements, such units could be used with great advantages for the fishing industry. Improvements which could easily be achieved would be:

- More suitable shape and size
- Improvements on hygienic standards by applying sealants, lining and improved cleaning
- Improved production methods
- Increased strength

The only traditional material for which accurate information is available is wood.

In many places boxes and containers made from local materials are readily available in sufficient quantities and at reasonable prices, while modern fish boxes/baskets are costly and difficult to find. This means that if hygienic standards and other practical aspects could be improved on, the fishing industry would have less problems.

Spoilage Due to Lack of Adequate Packing and Handling Facilities

The rate of spoilage and quality deterioration is often quite high in most developing countries.

Spoilage losses range from 15% up to more than 30% and the two most significant contributing factors are poor handling and inadequate chilling.

Use of suitable boxes/baskets during the handling process and special containers for transport and storage will undoubtedly contribute to improved quality and reduced spoilage.

Improved fish handling methods should involve:

- Introduction of boxes/containers onboard vessels (if no other built in system exists), in the receiving system, and in the distribution chain
- Improved hygienic standards (proper cleaning and desinfection of boxes/containers)
- Use of ice or other cooling methods for fish stored in boxes/containers
- Improvements and standardization of boxes and containers in use

Introduction of Insulated Containers in Developing Countries

Insulated containers have a wide range of applications and come in various types from the more sophisticated types with refrigeration machinery to simple types used with ice.

The more advanced units are used in long distance transportation and international marketing. Generally it will be the transport distance and the capacity of the marketing industry which will decide the usefulness of such containers. For the less advanced fishing industry, simple custom made insulated containers for iced fish will offer an inexpensive solution if one wants to introduce improved fish handling at critical stages. The following areas will achieve the greatest advantages from introduction of adapted insulated containers:

Small vessels: It is possible to arrange built-in or removable insulated containers in most small vessels. Care should be taken not to upset stability with such installations.

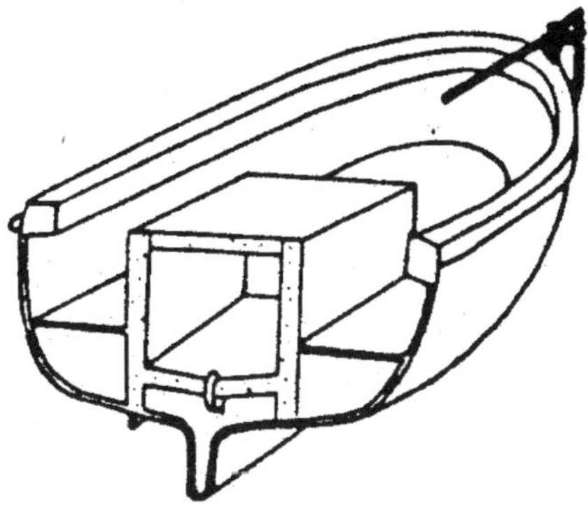

Fig. Insulated Hold in Small Vessel

Section of small fishing boat with insulated container for catch

Road Transport

Insulated containers are widely used in conjunction with trucks, trains and ships, but in many countries a significant amount of fish is transported over short distances on bicycles and handcarts and often in uninsulated boxes/ baskets. Light weight insulated containers can easily be made to suit such vehicles and would introduce a possibility for improvement in fish quality.

Retail Market

Retail sale of fresh fish requires space for display and storage. Storage can be arranged by use of insulated containers with ice. This will offer an inexpensive alternative which can keep fish in excellent condition for more than a week.

Such equipment will be within financial reach of small-scale retail businessmen who cannot afford any other refrigerated equipment.

Simple containers used for iced storage must have a smooth accessible interior that can be easily cleaned.

The lid should also be insulated, well sealed and positioned in the top. It is important that the container has a drainage opening at the lowest point of bottom. 5-10 cm of styrofoam insulation material will give sufficient protection and GRP (Glass Reinforced Plastic), various types of plastic sheeting, aluminium, galvanized steel and wood (marine plywood and timber) can be used as construction material. It is important that insulation material is distributed evenly through all parts of the box and that the inside which is going to be in direct contact with the fish is made, from hygienic and non-toxic material.

THE EFFECT OF HYGIENE DURING HANDLING

ONBOARD HANDLING

Much emphasis has been placed on hygienic handling of the fish from the moment of catching in order to ensure good quality and long storage life. The importance of hygiene during handling onboard has been tested in a series of experiments where various hygienic measures were employed .

The quality and storage life of completely aseptically treated fish (aseptic handling) were compared with fish iced in clean plastic boxes with clean ice (clean handling) and with fish treated badly, *i.e.*, iced in old, dirty wooden boxes (normal handling).

As expected, a considerable difference is found in the bacterial contamination of the three batches. However, a similar difference in the organoleptic quality is not detected. During the first week of storage no difference whatsoever is found. Only during the second week does the initial contamination level become important and the heavily contaminated fish have a reduction in storage life of a few days compared with the other samples.

On the basis of these data it seems sensible to advocate reasonably hygienic handling procedures including use of clean fish boxes. Very strict hygienic measures do not seem to have great importance. In comparison with the impact of quick and effective chilling, the importance of hygiene is minor.

The above-named observations have influenced the discussion about the design of fish boxes. Normally, fish are iced in boxes stacked on top of each other. In this connection it has been argued that fish boxes should have a construction that prevents the ice melt-water from one box draining into the box underneath it. In a system like this, some bacterial contamination of fish in the bottom boxes would be avoided, as melt-water usually contains a large number of bacteria. However, practical experience as well as experiments have shown that this type of contamination is unimportant, and it may be concluded that fish boxes allowing the drainage of melt-water from upper into lower boxes are advantageous because the chilling becomes more effective.

Inhibition or Reduction of the Naturally Occurring Microflora

In spite of the relatively minor importance of the naturally occurring microflora in the quality of the fish, much effort has been put into reduction or inhibition of this microflora. Many of these methods are only of academic interest. Among these are (at least until now) attempts to prolong the storage life by using radioactive irradiation. Doses of 100 000 - 200 000 rad are sufficient to reduce the number of bacteria and prolong storage life, but the process is costly and, to many people, unacceptable in connection with human food.

Another method which has been rejected because of concern about public health is treatment with antibiotics incorporated in the ice.

A method that has been used with some success over recent years is treatment with CO_2, which can be applied either in containers with chilled seawater or as part of a modified atmosphere during distribution or in retail packages.

It should also be mentioned that washing with chlorinated water has been tried as a means of decontaminating fish. However, the amount of chlorine necessary to prolong the storage life creates off- flavours in the fish meat . The newly-caught fish should be washed in clean seawater without any additives. The purpose of the washing is mainly to remove visible blood and dirt, and it does not cause any significant reduction in the number of bacteria and has no effect on storage life.

The Effect of Anaerobic Conditions and Carbon Dioxide

High CO_2 concentrations can reduce microbial growth and may therefore extend the shelf life of food products, where spoilage is caused by microbial activity. Technological aspects of modified atmosphere packaging (MAP) have since been studied. Today, materials and techniques for storage of bulk or retail packed foods are available.

The effect of anaerobic conditions and modified atmospheres on the shelf life of fish products. The safety aspects are reviewed in Farber (1991) and Reddy *et al.* (1992).

Effect on Microbial Spoilage

Vacuum packaging (VP) and MAP, with high CO_2 levels (25% - 100%), extends the shelf life of meat products by several weeks or months.

Table : Effect of packaging on the shelf life of chilled fish and meat products

Type of product	*Storage temp.*	*Shelf life (weeks)*		
		Air	*VP*[a]	*MAP*[b]
Meat (beef, pork, poultry)	1.0 - 4.4°C	1 - 3	1 - 12	3 - 21
Lean fish (cod, pollock, rockfish, trevally)	0.0 - 4.0°C	1 - 2	1 - 2	1 - 3
Fatty fish (herring, salmon, trout)	0.0 - 4.0°C	1 - 2	1 - 2	1 - 3
Shellfish (crabs, scampi, scallops)	0.0 - 4.0°C	½ - 2	-	½ - 3
Warmwater fish (sheepshead, swordfish, tilapia)	2.0 - 4.0°C	½ - 2	-	2 - 4

a) VP: Vacuum packed

b) MAP: Modified atmosphere packed (High CO_2 concentrations (25 - 100%)

In contrast, the shelf life of fresh fish is not affected by VP and only a small increase in shelf life can be obtained by MAP. Differences in spoilage microflora. and in pH are mainly responsible for the observed differences in the shelf life of fish and meat products. Spoilage of meat under aerobic conditions is caused by strict aerobic Gram-negative organisms, primarily *Pseudomonas spp.* These organisms are strongly inhibited by anaerobic conditions and by CO_2. Consequently, they do not play any role in the spoilage of packed meat. Instead the microflora, of VP and MAP meat products changes to be dominated by Gram-positive organisms (Lactic Acid Bacteria), which are much more resistant to CO_2 (Molin, 1983; Dainty and Mackey, 1992). Fish stored under aerobic conditions are also spoiled by Gram negative-organisms, primarily *Shewanella putrefaciens.*

The spoilage flora on some packed fish products was found to be dominated by Grampositive microorganisms and in this way the microflora, was similar to the flora on packed meats; see Stammen *et al.* (1990) for a review. For packed cod, however, the Gram-negative organism *Photobacterium phosphoreum* has been identified as the organism responsible for spoilage. The growth rate of this organism is increased under anaerobic conditions and this may explain the importance of the organism in VP cod. In CO_2-packed fish, the growth of *Shewanella putrefaciens* and of many other microorganisms found on live fish is strongly inhibited. In contrast P. *phosphoreum* was shown to be highly resistent to CO_2.

It was also shown that the limited effect of CO_2 on growth of this bacteria correspond very well with the limited effect of CO_2 on the shelf life of packed fresh cod. P. *phosphoreum* reduces TMAO to TMA while very little H_2S is produced during growth in fish substrates. Spoiled VP and MAP cod is characterized by high levels of TMA, but little or no development of the putrid or H_2S odours typical for some aerobically stored spoiled fish. The growth characteristics of P. *phosphoreum* and the metabolic activity of the organism thus explain both the short shelf life and the spoilage pattern of packed cod .

The shelf life of VP and MAP cod is similar to various other sea food products. P. *phosphoreum* is widespread in the marine environment and it seems likely that this organism or other highly CO_2 resistent microorganisms are responsible for spoilage of packed sea food products. The best effect of MAP storage on shelf life has been obtained with fish from warm waters. The shelf life of these products, however, is still relatively short compared to meat products. Very low bacterial level (10^5-10^6 cfu/g) has been found at the time of sensory rejection of some packed fish products. In these cases non-microbial reactions may have been responsible for spoilage.

Effect of Non-microbial Spoilage Reactions

CO_2 is dissolved in the water phase of the flesh of MAP fish and a decrease in pH of about 0.2- 0.3 units is observed, depending on the CO_2 concentration in the surrounding gaseous atmosphere. The water-holding capacity of muscle proteins is decreased by decreased pH and an increased drip loss is expected for fish stored in high CO_2 concentrations. Increased drip has been found for cod fillets, red hake, salmon, and shrimps but not for herring, red snapper, trevally, Dungeness crab, and rockfish.

Coyne (1933) and many later studies have found the textural quality of fish stored in 100% CO_2 to be reduced. However, up to 60% CO_2 has no negative effect on the texture of cod. The colour of the belly flaps, of cornea, and of the skin may be altered for whole fish stored in high CO_2 concentrations . Packaging may also stimulate the formation of metmyoglobulin in red- fleshed fish and thereby result in a darkening of fish muscles. Although oxygen-containing modified atmospheres have been used, the development of rancid off-odours in fatty fish species has not been registered as a problem .

Carbon Dioxide used in Combination with Refrigerated Seawater Systems

Only the effect of addition of CO_2 to RSW will be considered in this section. Table shows the effect of RSW and RSW + CO_2 on the shelf life of various fish products, as compared to storage in ice. An evident shelf life-extending effect of CO_2 is only seen with some species. Several negative effects of adding CO_2 to RSW-systems have been observed. The fish colour and texture were negatively influenced, and CO_2 dissolved in the flesh made mackerel unsuitable for canning.

CO_2 acidifies the seawater, and a lowered pH inhibits the enzymatic reactions that otherwise lead to black spots in shrimps and prawns. The shelf life of pink shrimps can be more than doubled by storage in RSW + CO_2, where, compared to ice storage, colour, texture, flavour, and odour were improved. RSW+CO_2 stored prawns, however, may be unacceptably tough and have a "soft shell" appearance .

Future Application of Carbon Dioxide for Shelf Life Extension

For most MAP seafoods, the production of TMA is delayed by only a few days compared to aerobic or anaerobic storage. This indicates that fish products in general are contaminated with a highly CO_2 resistent microflora of TMAO reducing organisms. Very high CO_2 concentrations can inhibit microbial growth but high levels of CO_2 have a negative effects on other aspects of the fish quality. MAP has found little practical application with fish products as compared to meat products. The main reasons for this are probably that:

- MAP used with retail packs is an expensive technique
- the prime fish quality is not improved
- only small shelf life extensions are obtained
- MAP cannot replace good chilling or good hygienic production conditions

- toxin production of *Clostridium botulinum* is increased for bacteria growing under anaerobic conditions, and this may be of importance for the safety of packed fish.

Packaging, however, can be used simply because packed products are more convenient to handle, *e.g.*, in supermarkets. According to the EEC Council Directive of 22 *July* 1991 (91/493/EEC), VP and MAP fish products are considered as fresh products. Consequently, CO_2 can be used for preservation of fresh fish products, when a shelf life extension of only a few days is found to be sufficient.

The negative effect Of CO_2 on fish colour is primarily a problem for whole fish and the negative effect of CO_2 on texture and drip loss is only observed with high CO_2 concentrations. A pronounced effect on growth of *S. putrefaciens* and on many other bacteria is obtained with even moderate CO_2 concentrations (40-80%). It is therefore likely that, in the future, MAP will be used in combination with preservation techniques that has been developed specifically to inhibit growth of CO_2 resistent TMAO reducing marine spoilage bacteria such as P. *phosphoreum*. The effect of MAP also seems to depend on fish species and further studies are needed to determine if MAP can give interesting shelf life extensions for other fish species, *e.g.*, those from warm waters. Finally, high CO_2 concentrations could be used for fish intended for fishmeal as the negative effects of CO_2 on colour and texture in this case are less important.

The Effect of Gutting

It is a common experience that the quality and storage life of many fish decrease if they have not been gutted. During feeding periods the fish contain many bacteria in the digestive system and strong digestive enzymes are produced. The latter will be able to cause a violent autolysis *post mortem,* which may give rise to strong off-flavour especially in the belly area, or even cause belly-burst. On the other hand, gutting means exposing the belly area and cut surfaces to the air thereby rendering them more susceptible to oxidation and discoloration. Thus, many factors such as the age of the fish, the species, amount of lipid, catching ground and method, etc., should be taken into consideration before deciding whether or not gutting is advantageous.

Fatty Species

In most cases,small- and medium-sized fatty fish such as herring, sardines and mackerel are not eviscerated immediately after catch. The reason for this is partly that a large number of small fish are caught at the same time and partly because of problems with discoloration and the acceleration of rancidity.

However, problems may arise with ungutted fish during periods of heavy feeding due to belly- burst. The reactions leading to belly-burst are complex and not fully understood. It is known that the strength of the connective tissue is decreased during these periods and that post mortem pH is normally lower in well-fed fish, this also weakens the connective tissue. Furthermore, it seems that the type of feed ingested may play an important role in the belly-burst phenomenon.

Lean Species

In most North European countries, the gutting of lean species is compulsory. It is based on the assumption that the quality of these species suffers if they are not gutted. In the case of cod, it has been shown that omission causes a considerable quality loss and a reduction in the storage life of five or six days. After only two days from catch, discoloration of the belly area is visible and the raw fillet acquires an offensive cabbagey odour.

These volatile, foul-smelling compounds are mostly found in the gut and surrounding area whereas the amount of volatile acids and bases is relatively low in the fillet itself. These chemical parameters are, therefore, not useful for distinguishing between gutted and ungutted fish .

In the case of haddock *(Melanogrammus aeglefinus),* whiting *(Merlangius merlangus,* saithe *(Pollachius virens)* and blue whiting *(Micromesistius poutassou),* it is observed that ungutted fish stored at 0°C suffer a quality loss compared with gutted fish, but the degree varies. Some off-odours and off-flavours are detected, but ungutted haddock, whiting and saithe are still acceptable as raw material for frozen fillets after nearly one week on ice . Quite different results are obtained with South American hake *(Merluccius gayi),* where no difference is observed between gutted and ungutted fish .

THE EFFECT OF FISH SPECIES, FISHING GROUND AND SEASON

INFLUENCE OF HANDLING, SIZE, PH, SKIN PROPERTIES

The spoilage rate and shelf life of fish is affected by many parameters and, fish spoil at different rates. In general it can be stated that larger fish spoil more slowly than small fish, flat fish keep better than round fish, lean fish keep longer than fatty fish under aerobic storage and bony fish are edible longer than cartilaginous fish. Several factors probably contribute to these differences and whereas some are clear, many are still on the level of hypotheses.

Table : Intrinsic factors affecting spoilage rate of fish species stored in ice

Factors affecting spoilage rate	*Relative spoilage rate*	
	fast	*slow*
size	small fish	larger fish
post mortem pH	high pH	low pH
fat content	fatty species	lean species
skin properties	thin skin	thick skin

Rough handling will, result in a faster spoilage rate. This is due to the physical damage to the fish, resulting in easy access for enzymes and spoilage bacteria. The surface/volume ratio of larger fish is lower than that of smaller fish, and, as bacteria are found on the outside, this is probably the reason for the longer shelf life of the former. This is true within a species but may not be universally so.

Post mortem pH varies between species but is, higher than in warm- blooded animals. The long rigor period and the corresponding low pH (5.4-5.6) of the very large flatfish, halibut (*Hippoglossus hipoglossus*), has been offered as an explanation for its relatively long iced storage life. However, mackerel will often also experience a low pH and this seems to have little effect on shelf life.

The skin of the fatty pelagic fish is often very thin, and this may contribute to the faster spoilage rate. This allows enzymes and bacteria to penetrate more quickly. On the contrary, the thick skin of flatfish and the antibacterial compounds found in the slime of these fish may also contribute to the keepability of flatfish.

As described earlier, the slime of flat fish contains bacteriolytic enzymes, antibodies and various other antibacterial substances. Although large differences exist in the content of TMAO, this does not seem to affect the shelf life of aerobically-stored fish but rather the chemical spoilage profile of the species.

Table : Shelf life of various fish species from temperate and tropical waters.

Species	*Fish type*	*Shelf life (days in ice)*	
		temperate	*tropical*
Marine species		2-24	6-35
cod, haddock	lean	9-15	
whiting	lean	7-9	
hake	lean	7-15	
bream	lean/low fat		10-31
croaker	lean		8-22
snapper	lean		10-28
grouper	lean		6-28
catfish	lean		16-19
pandora	lean		8-21
jobfish	lean		16-35
spadefish	lean/low fat		21-26
batfish	lean		21-24
sole, plaice,	flat	7-21	21
flounder	flat	7-18	
halibut	flat	21-24	
mackerel[1)]	high/ low fat	4-19	14-18
summer herring	high fat	2-6	

winter herring	low fat	7-12	
sardine	high fat	3-8	9-16
Freshwater species		9-17	6-40
catfish	lean	12-13	15-27
trout	low fat	9-11	16-24
perch	lean/ low fat	8-17	13-32
tilapia	lean		10-27
mullet	lean		12-26
carp	lean/ low fat		16-21
lungfish	lean/ low fat		11-25
Haplochromis	lean		6
shad	medium fat		25
corvina	medium fat		30
bagré	medium fat		25
chincuna	fatty		40
pacu	fatty		40

1) fat content and shelf life subject to seasonal variation.

In general, the slower spoilage of some fish species has been attributed to a slower bacterial growth, and Liston (1980) stated that "different spoilage rates seem to be related at least partly to the rate of increase of bacteria on them".

Influence of Water Temperature on Iced Shelf Life

Of all the factors affecting shelf life, most interest has focused on the possible difference in iced shelf life between fish caught in warm, tropical waters and fish caught in cold, temperate waters. In the mid- and late sixties it was reported that some tropical fish kept 20-30 days when stored in ice .

This is far longer than for most temperate species and several studies have been conducted assessing the shelf life of tropical species. Comparison of the data is, as pointed out by Lima dos Santos (1981), difficult as no clear definition has been given on a "tropical" fish species and as experiments have been carried out using different sensory and bacteriological analyses.

Several authors have concluded that fish taken from warm waters keep better than fish from temperate waters whereas Lima dos Santos (1981) concluded that also some temperate water fish species keep extremely well and that the longer shelf lives in general are found in fresh water fish species compared to marine species.

However, he also noted that shelf life of more than 3 weeks, which is often observed for fish caught in tropical waters, never occurs when fish from temperate waters are stored in ice.

The iced shelf life of marine fish from temperate waters varies from 2 to 21 days which does not differ significantly from the shelf life of temperate freshwater fish ranging from 9 to 20 days.

Contrary to this, fish caught in tropical marine waters keep for 12-35 days when stored in ice and tropical freshwater fish from 6 to 40 days. Although very wide variations occur, tropical fish species often have prolonged shelf lives when stored in ice. When comparisons are made, data on fatty fish like herring and mackerel should probably be omitted as spoilage is mainly due to oxidation. Several hypotheses have been launched trying to explain the often prolonged iced spoilage of tropical fish. Some authors have noted an absence in development of TMA and TVN during storage and suggested that the spoilage of tropical fish is not caused by bacteria .

The lack of development of TMA and TVN may be explained by a spoilage dominated by *Pseudomonas* spp.; however, qualitative bacteriological analyses must be carried out to confirm or reject this suggestion. Low bacterial counts have been claimed in some studies, but often inappropriate media have been used for the examination and too high incubation temperatures (30°C) have not allowed the psychrotrophic spoilage bacteria to grow on the agar plates.

Reviewing the existing literature on storage trials of tropical fish species leads to the conclusion that the overall sensory, chemical and bacteriological changes occurring during spoilage of tropical fish species are similar to those described for temperate species.

Psychrotrophic bacteria belonging to *Pseudomonas spp*. and *Shewanella putrefaciens* dominate the spoilage flora of iced stored fish. Differences exist, in the spoilage profile depending on the dominating bacterial species.

Shewanella spoilage is characterized by TMA and sulphides (H_2S) whereas the *Pseudomonas* spoilage is characterized by absence of these compounds and occurrence of sweet, rotten sulphydryl odours. As this is not typical of temperate, marine fish species which have been widely studied, this may explain the hypothesis that bacteria are not involved in the spoilage process of tropical fish.

Despite the different odour profiles, the level at which the offensive off-odours are detected sensorially is more or less the same. In model systems (sterile fish juice) 10^8-10^9 cfu/ml of both types of bacteria is the level at which spoilage is evident. The relatively high post*mortem* pH is one of the reasons for the relatively short shelf life of fresh fish as compared to, for instance, chill stored beef.

It has been suggested that tropical fish species, such as the halibut from temperate waters, reach a very low pH, and that this explains the longer shelf life. However, pH values of 6-7 have been found in the studies of tropical fish species where pH has been measured . As the differences in skin properties are believed to contribute to the longer shelf life of flatfish, it has been suggested that this factor explained the extended shelf lives. It is indeed true that fish from warm waters often have very thick skin, but no systematic investigation has been carried out on the skin properties. As spoilage of fish is caused by bacterial action, most hypotheses dealing with the long iced shelf life of tropical fish species have centred around differences in bacterial flora. Shewan (1977) attributed the long iced shelf lives to the lower number of psychrotrophs on tropical fish. However, in 1977 only a very limited number of studies of the bacterial flora on tropical fish were published. During the last 10- 15 years several investigations have concluded that Gram-negative rod-shaped bacteria (*e.g., Pseudomonas, Moraxella* and *Acinetobacter)* dominate on many fish caught in tropical waters.

Similarly, Sieburth (1967) concluded that the composition of the bacterial flora in Narragansett Bay did not change during a 2-year survey even though the water temperature fluctuated with 23°C on a year-round basis. Gram (1989) showed that 40-90% of the bacteria found on Nile perch were able to grow at 7°C. The number of psychrotrophic bacteria is within one log unit of the total count, and the level of psychrotrophic organisms is not *per se low* enough to account for the extended iced storage lives of tropical fish; Jorgensen *et al.* (1989) showed that a two log difference in number of spoilage bacteria only resulted in a difference of 3 days in the shelf life of iced cod.

The bacterial flora on temperate water fish species resume growth immediately after the fish have been caught and rarely is a lag phase seen. Contrary to this, Gram (1989) concluded that a bacterial lag phase of 1-2 weeks is seen when tropical fish are stored in ice. Also, the subsequent growth of psychrotrophic bacteria is often slower on iced tropical than on iced temperate water fish. This is in agreement with Liston (1980) who attributed differences in shelf life to differences in bacterial growth rates.

Although a large part of the bacteria on tropical fish are capable of growth at chill temperatures, they will (as this has never been necessary) require a period of adaptation (*i.e.,* the lag phase and slow growth phase). Gram (1989) illustrated this by investigating the growth rate at 0°C of fish spoilage bacteria that had either been pre-cultured at 20°C or at 5°C. For some strains, the same bacterial strain would grow more quickly at 0°C if pre-cultured at 5°C than if pre-cultured at 20°C. Preculturing was done with several sub culture steps at each temperature. Similarly, Sieburth (1967) showed that although the taxonomic composition of the bacterial flora in Narrangansett Bay did not change with fluctuating temperature, the growth profile of the bacteria fluctuated following the water temperature. However, the adaptation hypothesis does not explain why some tropical fish spoil at rates comparable to temperate water fish.

It can be concluded that many factors affect shelf life of fish and that differences in the physiology of the bacterial flora are likely to be of major importance.

Off Flavours Related to Fishing Ground

Occasionally fish with off-flavours are caught, and in certain localities this is a fairly common phenomenon. Several of these off-flavours can be attributed to their feeding on different compounds or organisms. The planktonic mollusc, *Spiratella helicina,* gives rise to an off-flavour described as "mineral oil" or "petrol". It is caused by dimthyl-B-propiothetin which is converted to dimethylsulphide in the fish . The larvae of *Mytilus* spp. cause a bitter taste in herring. A very well known off-flavour is the muddy-earthy taint in many freshwater fish.

The flavour is mainly caused by two compounds: geosmin (1a, 10ß-dimethyl-9a-decalol) and 2-methylisoborneol, which also are part of the chemical profile of wine with cork flavour. Geosmin, the odour of which is detectable in concentrations of 0.01-0.1 μg/l, is produced by several bacterial taxa, notably the actinomycetes *Streptomyces* and *Actinomyces.*

An iodine-like flavour is found in some fish and shrimp species in the marine environment. This is caused by volatile bromophenolic compounds; and it has been suggested that the compounds are formed by marine algae, sponges and Bryozoa and become distributed through the food chain .

Oil taint may be found in the fish flesh in areas of the world where off-shore exploitation of oil is intensive or in areas where large oil spills occur. The fraction of the crude oil that is soluble in water is responsible for the off-flavours. This is caused by the accumulation of various hydrocarbon compounds, where particularly the aromatic compounds are strong flavourants .

6

Genetic Biotechnologies in Fishery Sector

MANAGEMENT OF GENETIC RESOURCES

This Background Document provides a summary of recently developed biotechnologies that could be used, or more widely used, in the fishery sector in developing countries. Genetic biotechnologies that can be used in fisheries and in aquaculture include those that help to manage genetic resources and those for genetic improvement. For management of genetic resources, markers can be used in the identification of management units and of endangered species to assist fishery management and they can also help broodstock management in stocking programmes. These markers may be genes, proteins (*i.e.* the products of genes), sequences of DNA or the phenotypic expression of genes (different colours, shapes etc.). In the 1960's, analysis of proteins revealed a wealth of genetic diversity in wild populations. Protein analysis is now relatively fast and inexpensive, but it requires tissue samples to be stored and transported frozen. DNA analysis is becoming the method of choice because of the small amount of tissue needed, the fact that the tissue can be stored dried or in alcohol, and because DNA analysis reveals much more genetic variation than protein analysis. Several kinds of DNA markers exist, such as RFLPs, AFLPs, RAPDs and microsatellites. These, as well as other kinds of markers, can be used to analyse gene frequencies and genetic variation in and between different groups of fish. Studies carried out using these technologies in fish populations have revealed high levels of genetic variation distributed throughout the fish genome.

GENETIC IMPROVEMENT TECHNOLOGIES

Genetic improvement technologies cover a range of techniques requiring different levels of expertise and resources. Chromosome-set manipulation (*i.e.* polyploidy induction) is an established technique to increase the number of chromosome-sets (ploidy number) in an organism. Temperature, chemical and pressure shocks applied to fish eggs can be used to produce triploid (3 chromosome-sets) individuals that have desirable culture traits. Sex-reversal and the production of single sex groups of fish is also a simple technology that combines hormone treatment and chromosome-set manipulation. Hybridisation, *i.e.* the mating of genetically different groups from the same species (intra-specific hybridisation) or from different species (inter-specific hybridisation), is a simple technique that is now easy to accomplish due to our increased knowledge of reproductive biology.

It can be used to combine good traits from two different species into one group of fish or to transfer a characteristic of one group to another. A problem is that breeding hybrids with hybrids results in a non-uniform and unpredictable group of fish that is generally not well suited for culture. Therefore, for hybrid production, the parent-lines must be maintained pure. The above genetic improvement techniques are considered short-term strategies, where the gains are seen in one or two generations. Selective breeding is a longer-term strategy where gains are accumulated at each generation of selection.

Molecular markers may now increase the efficiency of selective breeding by facilitating the identification of quantitative trait loci (QTLs), *i.e.* genes that control complex characters such as growth rate and environmental tolerance and, secondly, by making it possible to use molecular markers linked to QTLs to identify desirable individuals or families. Genetic engineering and the production of transgenic organisms is an active area of research and development in aquaculture. This is a medium-long term strategy in that development and testing of stable transgenic lines requires time. The large size and hardy nature of many fish eggs allows them to be manipulated rather easily and facilitates gene transfer by direct injection of a foreign gene or by electroporation, where an electric field assists gene transfer. We will briefly discuss currently available biotechnologies in the context of fishery management, aquaculture and conservation respectively.

BIOTECHNOLOGIES IN FISHERY MANAGEMENT

The role that the application of genetic principles can play in the sustainable use and conservation of living aquatic resources is being increasingly appreciated by resource managers, policy makers and the international community. Fishery management requires information on the fishery resources in order to be effective. Primary information needs include:

- an identification of the resource;
- the breeding or stock structure of the resource;
- an estimate of the size of the resource; and
- the identification of key habitat that the resource requires.

Genetic analysis of the resources can address these information needs. Gene and genotype frequencies of different markers can provide information on, inter alia, species identification, population stock structure, hybridisation and gene flow. Often, data from other sources, *e.g.* studies of tagged fish or of external characters of fish, cannot provide such information or are extremely difficult to collect in certain areas such as large river systems, floodplains or marine areas. The use of protein and DNA data in fishery management requires collection of baseline (or background) genetic information. Genetic data were used to determine how sub-groups of Pacific salmon differed from each other in the Pacific Northwest.

This required the analysis of hundreds of stocks of salmon but, once completed, endangered stocks were identified, levels of migration were estimated, and the contribution of different stocks to a mixed stock ocean fishery was estimated. Protein and DNA information has been used to identify endangered species that are either inadvertently captured in wild fisheries or that are purposefully taken illegally. DNA analysis of legally sold whale meat revealed that many samples came from protected species of whale and dolphin. Species of shark are often difficult to identify because it is only the fins or flesh that are for sale; DNA analysis can be used to identify the species that provided the tissue and has the added advantage that dried tissue or less than fresh samples from markets can be studied.

BIOTECHNOLOGIES IN AQUACULTURE

Genetic biotechnologies in aquaculture focus primarily on increasing growth rate, but also include disease resistance and increased environmental tolerance. There are several biotechnologies that can be applied to farmed aquatic species.

Selective breeding, *i.e.* traditional animal breeding, started with the common carp several thousand years ago. However, it has only recently been applied to a handful of other species of food fish such as catfish, trout and tilapia. Therefore, many farmed aquatic species are very similar to their wild relatives. Selective breeding programmes have yielded significant and consistent gains of 5-20% per generation in species of, inter alia, Atlantic salmon, catfish and tilapia. Hybridisation is a simple genetic technology that has become easier with the development of artificial breeding techniques, such as the use of pituitary gland extract and other hormones to initiate gamete development and induce spawning (*i.e.* the depositing of eggs), and an increased understanding of environmental cues that influence reproduction, such as day length, temperature or water current. Many of the natural reproductive isolating mechanisms that species develop in the wild can now be overcome by fish farmers.

These improvements in reproductive technologies have also assisted aquaculturists greatly in their efforts to domesticate aquatic species. In addition, by making it possible to remove the natural constraints and timing of breeding, farmers are able to mate many more species at the times that are most beneficial, and thus help to ensure a steady and consistent supply of fish to the market. Chromosome-set manipulation can be used to produce triploid organisms that generally do not channel energy into reproduction because of problems associated with development of reproductive organs. Initially it was thought that this energy saving would result in increased growth rate, but this seems not to be the case. The real advantage of triploids seems to be in their functional sterility.

For example, triploid oysters do not produce gonads (*i.e.* reproductive glands) and are therefore marketable at times of the year when mature oysters have an off-taste because of production of gametes (*i.e.* sex cells - the ovum, or egg (female), and sperm (male)). In aquaculture, one sex is often more desirable than the other. For example, female sturgeon produce caviar, male tilapia grow faster than females whereas it is the female trout and salmon that generally grow faster than the males. The production of single sex groups of fish takes advantage of these differences between the sexes and can be accomplished by manipulation of the developing gametes and embryo.

The manipulation can be in the form of denaturing (*i.e.* destroying) the DNA in gametes followed by chromosome-set manipulation or by hormonal sex-reversal and subsequent breeding. The phenotypic sex of many aquatic species can be changed by administering appropriate hormones.

For example, genetically male tilapia can be turned into females through estrogen treatments. These genetic males when mated with normal males produce a group of all-male tilapia that grow faster and have less unwanted matings (that lead to overcrowding and stunting) than a group of mixed-sex tilapia. Some of the all-male offspring would have two male chromosomes and these could be used as broodstock for subsequent generations, thus avoiding the use of hormones in the broodstock. Hybridisation can also be used to produce single sex groups of fish, when the sex-determining mechanisms in the parental lines are different (for example, hybridisation of Nile tilapia and the blue tilapia). Genetic engineering is a vague term that has come to be nearly synonymous with gene transfer *i.e.* the production of transgenic fish or genetically modified organisms (GMOs). This technology is progressing rapidly and it is now possible to move genes between distantly related species. Gene transfer in fish has usually involved genes that produce growth hormone and has been shown to dramatically increase growth rate in carp, catfish, salmon, tilapia, mudloach and trout.

A gene from the winter flounder that produces an anti-freeze protein was put into salmon in the hope of extending the farming range of the fish.

The gene did not produce enough of the protein to extend the salmon's range into colder waters, but it did allow the salmon to continue growing during cold months when non-transgenic salmon would not grow. Transgenic technology is currently in the research and development stage. To our knowledge there are no transgenic aquatic plants or animals available to the consumer.

Cryopreservation

The development of cryopreservation or low-temperature technology allows the short- and long-term storage of gametes. Currently, these low-temperature techniques can only be used on male gametes; eggs and embryos can generally not be stored in this way. Freezing gametes can increase the flexibility of a fish breeder, especially when breeding species where the sexes mature or migrate at different times, when the breeding season is very short, when the breeders are far apart or when one sex is exceptionally rare.

Fish Health

Genetic biotechnologies are being used to improve fish health through conventional selection for disease resistance and through the use of molecular investigation of pathogens for characterisation and diagnosis.

DNA-based technologies are being used now to characterise different species and strains of pathogens. Genetic characterisation of the pathogen may also reveal information about its origin, *e.g.* DNA analysis revealed two strains of crayfish plague fungus in Sweden: one from the local species and one originating in Turkey. Once the pathogen is characterised, DNA probes can be developed to screen for specific pathogens in tissue, whole animals and even in water and soil samples. These techniques are being used to detect viral diseases of marine shrimp throughout the world and for bacterial and fungal pathogens in fishes in many areas.

Genetically engineered vaccines are also being developed to protect fish against pathogens. Genetic immunisation of rainbow trout with a glycoprotein gene from the virus causing viral haemorrhagic septicaemia has recently been shown to induce high levels of protection against the virus. Work is also underway on immunising carp, salmon and other fishes with genetically engineered vaccines for other diseases. The new molecular techniques are extremely sensitive and can identify pathogens in fish long before there are any clinical signs of the disease. This has implications for quarantine and the trade of aquatic species, which is currently governed by the World Trade Organisation and the Office International des Epizooties. Trade can be restricted based on the disease status of a product or a region; identification of minute quantities of a pathogen or of a new strain of an existing pathogen could change or influence existing trade patterns.

Farming Systems

Farming systems for aquatic species are diverse and include industrial scale farms, family ponds and culture based fisheries (stocking), in both developed and developing countries. Often, there is a division of the production process where fingerlings (*i.e.* small fish, especially up to one year of age) or eggs are produced by the seed-supplier, but the grow-out to market size is done elsewhere.

In the case of sea going salmon, there is often a seed supplier operating a hatchery near a river, a fingerling producer in a freshwater lake, and another group that grows the fish to market size in the sea. Marine shrimp hatcheries in Asia are usually small family owned ventures, whereas in Latin America they are more industrial in scale. Appropriateness of genetic biotechnologies must take these different systems into consideration.

Biotechnologies in Conservation

Genetic biotechnologies can be used to reduce the impacts of farmed fish on wild populations, to identify and manage endangered species and to manage captive populations in aquaria or in species recovery programmes. In several areas, farmed fish must be made triploid, *i.e.* sterile, in order to reduce their impact on wild populations should they escape from the fish farm.

Generally, the planned use of transgenic fish also includes the provision that they are sterile, to reduce the chance of mixing with other fishes. Genetic manipulation and polyploidisation can be combined to regenerate endangered species. This can be done from frozen sperm by denaturing the DNA in an egg of a related species, fertilising with frozen sperm from the endangered species and then duplicating the chromosome-set of the fertilised egg.

BIOTECHNOLOGY IN FISHERIES AND AQUACULTURE

The fisheries and aquaculture industries in Asia contribute about 45 percent of world fish production . These industries are significant contributors to the food supply, livelihood, foreign exchange earnings and socio-economic stability in rural areas of several Asian countries. At the global level fish production is stagnant, at around 100 million tonnes, or has even slightly declined in recent years. Overharvesting of the world's oceans is recognized as a global threat. To obviate this threat, the production of cultured fish must be accelerated. The ability to produce transgenic fish and shellfish in culture, which grow faster and larger with more efficient utilization of nutrients, is of particular value to developing countries, not only as a source of food, but also as export products.

Biotechnology offers great promises for fish disease control and feed production as well. Gynogenetic and polyploid progeny production and the use of growth hormones have significantly contributed to fish production . The application of hypophysation (injecting of pituitary gland extracts into mature fishes for the induction of spawning) has been widely practiced in Asia during the past two decades. The use of purified gonadotrophin hormone synthetic luteinizing hormone-releasing hormone (LH-RH) has further improved induced spawning techniques.

In China, following detailed discussions and workshops in the 1980s, a policy and plan for aquaculture biotechnology were included in the Seventh Five-Year Plan (1985-90). In 1987, they were also listed in the "863 Programmes", the state development programmes on high technologies, and received great attention from the Ministry of Aquaculture. Fish cell culture, sex control, chromosome manipulation, and to some extent transgenic production, have been the main areas of work.

China's success in fish cell culture, of grass carp, was reported in 1979 by the Freshwater Fishery Research Institute of Zhejiang Province. Up to now, more than ten fish cell strains have been cultured for producing vaccines. Bioreactors for industrial fish cell culture are being developed. Seaweed cell culture in China was first reported in 1985 and is being developed for further breeding work. Diploid gynogenesis, coupled with sex control is being used for mass production of all female common and crucian carps. Although triploid hybrids between 4n common carp (male) and 2n grass carp (female) have been obtained, the work is confined to laboratories.

China was one of the pioneers in transplanting of fish cell nucleus. In the mid-1970s, the reciprocal transplant of nucleus was accomplished between common carp and crucian carp. Of the cell hybrid fishes, 30 grew to sexual maturity and produced F2 and F3 progenies. In 1984, the F3 progenies were mass produced and these showed stable heredity and had 30 percent faster growth than their grandparents. A similar practice was carried out on grass carp and Megalobrama amblycephala; 26 of the hybrids grew well, matured sexually and were mass produced by aquaculturists. In 1988, live fries were produced from the fusion of blastulla cells and eggs in different fishes: Paramisgrunus dacryanus, Megalobrama amblycephala, grass carp, common carp and red crucian carp. This technique holds great promise for fish breeding.

In fishes, the two sexes differ in their productivity and body weight. For instance, in Tilapia males grow faster than the females, the converse of which is true for carps. Of the several methods of sex manipulation, through distant hybrids, hormone treatment, induced diploid gynogenesis, induced triploids, antoimmure castration, surgical castration, and gene technique, in China the first two are commonly used. Hybrid populations between Tilapia nilotica (female) and T. aurea (male) were 95 to 98 percent male, and on an average yielded about 25 percent higher than the higher yielding parent (T. nilotica). Moreover, the hybrid fish could tolerate lower water temperatures, and could be harvested more easily. Such all-male populations are already being mass cultured in China, the Philippines, Thailand and other countries. Israel has made comprehensive use of this technique.

As regards the use of hormones, using steroid hormones, induced females of T. mossambicais when crossed with male T. nilotica, produced super males with a "yy" sex chromosome combination. The "super males" crossed with female T. nilotica produced 100 percent male progenies which possessed the twin superiorities of being hybrid as well as male. The hybrids grew 38.5 percent faster than T. nilotica and the mass production increased by 43.4 percent. Although a sizeable gain, the difficulty in identifying the "super male" fish at fry stage is a major constraint to widespread exploitation.

As regards genetic engineering, Chinese scientists have succeeded in identifying and isolating the gene controlling growth hormone (GH) from salmon, common carp, grass carp and silver carp. Transgenic carps carrying human growth hormone gene have been produced. Antifreeze gene has also been identified and used for improving the antifreeze ability of Tilapia. Successful cryopreservation of fish sperm, involving eight species has been established in China.

India, with a coastline of 7 512 km has 2.02 million km^2 of Exclusive Economic Zones, 4.7 million hectares of identified freshwater areas, and 1.7 million hectares of brackish-water areas. Its annual production of fish is around 4 million tonnes, half of it from aquaculture, and there is ample scope for enhancing fish production in the country. To supplement the fish feed resources, the use of bacterial cells as larval fish feed and processed aquatic plants as grow-out feed has been tested with success in India .

In Japan, chromosomal manipulation and modern biotechnology are being widely applied to improved fish production. Gynogenesis, artificial sex reversal and induced polyploidy are being routinely used for increasing productivity as well as the quality of fish.

The growth hormone of salmon and yellowtail is being mass produced through transgenic bacteria and used as a growth promoter on other fish such as rainbow trout. In the field of transgenic fish, the introduction and manifestation of the crystalline genes of chickens into killifish embryos were successfully achieved in 1986 at Kyoto University . Other useful genes are being identified and introduced into commonly used fish species.

New vaccines are also being developed to treat fish diseases. As regards fish feed, the development of microparticulate diets for larval fish to replace live foods in hatcheries is a major achievement. Mass production of ayu and carp fry in Japan with such diet formulates has been successful .

Malaysia has a relatively long coastline, stretching for about 4 334 km and its marine resources contribute about 9.5 percent of the total value added in the agricultural sector; production has been increasing at a high annual growth rate of about 8 percent over the past decade.

Moreover, the industry employs about 120 000 fishermen. The country clearly sees the scope for the application of biotechnology in fish production through genetic improvement of cultured fish, disease resistance and improved growth rate. Yet there is a gap in absorption or diffusion of the proven technologies.

Besides fish, marine algae are now recognized as important resources for the production of valuable chemicals in Malaysia and local species of marine macroalgae (seaweed) are being screened for valuable chemicals. Some of the species studied and their products include: Gracilaria changii for agar and Turbinaria conoides, Sargassum baccularia and Sargassum siliquosum for algiric acid and antimicrobial compounds.

Singapore considers its marine bioresources as one of the most important and the government is highly supportive of the marine biotechnology venture. The National University of Singapore (NUS) is playing a leading role in this direction. Its Bioprocessing Technology Unit, established in 1990, concentrates on two main areas: fermentation and cell culture; and the isolation, separation and production purification technology.

In addition, the Unit also supplies to specific users, especially synthesized antibodies for immunobiological assays and analyses, and custom-made digonucleotide primers for nucleic acid research. In 1992, the NUS and the Economic Development Board (EDB) established a Bioscience Centre in the University, with the following objectives: (i) to provide a multiuser central facility for biotechnology, (ii) to serve as a vehicle for collaborating with industries and other institutions in Singapore or abroad, in areas of common interest, and (iii) to provide training in biotechnology.

The areas of research of the Centre are: bioactive compounds from marine organisms; toxins and venoms research; plant bioactive products; and products from microorganisms. Other related new establishments are the Centre for Natural Product Research and the Centre for Medical Informatics. The Reef Ecology Study Team of the Department of Zoology in NUS is actively surveying the distribution of the various reef organisms and the Bioscience Centre is screening organisms for antifouling, antibacterial, antitumour, antiviral, antifungal, anti-inflammatory and insecticidal compounds. Antibacterial compounds are being isolated and characterized from horseshoe crabs. Biological properties and potential therapeutic application of toxins and other bioactive compounds in two marine species, namely the "Thunder Crab" (Lyphozozymus pictor) and the stonefish (Sysanceja horrida), have been studied.

In addition, a few bioactive compounds have been purified from five local sea anemone species. These include a cytolysin, a trypsin inhibitor and two putative neurotoxins. Transgenics are also being attempted in a few fish species such as Marble Goby, grouper and salmon. The Marble Goby growth hormone gene has been cloned. NUS scientists are now in the process of combining this gene with a heterologous promoter to make a gene construct for transgenesis. Aquaculturists have been advised of the applications of (i) thyroid hormones in enhancing embryo and fry development; and (ii) gonado-tropin releasing hormones to help induce yolk formation and spawning. The hormones are being encapsulated into synthetic capsules for commercial use. Immunologists are studying the immune responses in fish and are developing various disease control methods, including vaccination with attenuated disease agents to develop antibodies.

Singapore is one of the major exporters of tropical aquarium fishes in the world. The DAN fingerprinting technology is being used to identify proper breeder stocks to be used in breeding programmes to maintain the quality and health of the fishes.

Aquaculture is most developed in Thailand. A Marine Biotechnology Unit (MBU) was established jointly by the National Centre for Genetic Engineering and Biotechnology and Aquatic Resources Research Institute of Chulalong Korn University in 1987. The Unit has been screening algae strains for feed purposes. Using cloned mitochondrial DNA, RFLP patterns of giant freshwater prawns (Macrobrachium rosenberghi) were studied for analysing variation patterns to design appropriate breeding plans. Studies are in progress to manipulate the Gonad Inhibiting Hormones (GIH) for regulating the growth, maturity and reproductive success of prawns.

To control the most common and yield-reducing disease, vibriosis, caused by the bacteria Vibrio anguillarum, of the widely cultured fish seabass, vaccines have been developed. Chromosome manipulation of oyster has also been standardized. Studies on the gametogenic cycle of Thai abalone (Haliotis ovina) are proving helpful for resource regulation and management, and for broadstock conditioning, induction of spawning, control of mating and seed production for land-based production. An outdoor mass culture of benthic diatom, Navicula sp., as a food source of abalone spot was also standardized recently.

Research on marine natural products in Thailand began a few years ago and provides valuable information on the structures and potential applications of the compounds and their derivatives from Thai sponges. Such information could be, apart from the bioactivities of the compounds that can possibly be applied further to pharmaceutical use, helpful in the chemotaxonomy of Thai sponges. The use of seaweeds as a source of phycocolloids is being promoted. The Biopolymer Research Unit (BU) of the Srinakarinwirot University, established in 1987, has developed a simple and low-cost technique for agar extraction from Gracelaria sp. seaweeds which could be applied at the village level. By developing and using chitin filter medium, the technique has improved the quality of the agar produced. The chitin could also be changed to chitosan, another valued product that is biodegradable and nonpolluting and can be used as a binder in recycling paper. This technology changes the shrimp wastes into more value-added products. The Unit is currently operating a project entitled "Technology Development of Shrimp Shell Processing". Other work relates to the development of biomembranes, copolymer membrane and biobeads. Another value-added product is a fluorescence pigment of R-phycoerythrin for protein markers.

The Marine Biotechnology Centre at Burapha University is concentrating on population dynamic studies of phytoplankton and zooplankton and identification of new pharmaceuticals from marine organisms. From the various species screened, the Centre found that 21 marine species had shown some degree of antimicrobial action.

Twenty-three extracts from seven species had shown positive results when subjected to haemolytic testing against erythrocytes of 21 species. From the above, it can be seen that the following biotechnology-related research work on marine organisms is in progress: studies on bioactive metabolites from Thai sponges, isolation of pure culture of some marine microalgae, isolation of pure culture of some marine bacteria, and the use of microalgae in biological control of marine animal diseases.

While considerable progress has been made in fish biotechnology in Asia, the following aspects need to be further strengthened :

- Genetic improvement of cultured fish for improved growth, disease resistance and adaptability to new farming systems.
- Development of low-cost but efficient grow-out diets for cultured fishes using locally available feedstuffs and microbiological processes.
- Utilization of detailed and microbial feed sources for pond cultured fish.
- Production of vaccines and monoclonal antibodies for the prevention and control of bacterial and virus diseases.
- Development of drugs and other biotechnological products from marine organisms.
- Improved availability of mother stocks and seeds, especially of brackish water marine fish and shellfish.

AQUACULTURE AND MARINE BIOTECHNOLOGY

MAIN THRUST OF ACTIVITIES

* Development of recombinant diagnostics and vaccines for major diseases in aquaculture
* Molecular biology of Indian species, identification of useful genes and development of indigenous gene constructs for fish transgenesis work
* Genomics & proteomics studies in aquaculture species
* Bioactive molecules from aquatic organisms for therapeutic and industrial applications
* In vitro tissue culture, cell culture system development in various aquaculture species
* Front-line demonstrations to prove techno-economic viability of seed production and aquaculture in non-traditional species

* Programmes on new feed development, nutrition, breeding, spawning agents, health and post harvest aspects etc. for enhancing aquaculture productivity
* Programme support in marine biotechnology
* Training Programme in Molecular Biology for Fisheries Scientists

Achievements

Diagnostics and vaccine development in aquaculture are a major priority apart from development of suitable cell culture systems, immunostimulants, anti-bacterial and anti-cancerous agents. Studies have been supported on isolation of novel enzymes from deep-sea fungi, molecular cloning and gene expression for alkaline protease and molecular epidemiology of white spot shrimp disease in brackish water system. DNA vaccines developed against Aeromonas are being studied for the efficacy and challenge experiments. Immunodot-based diagnostic kit is being used for detection of white spot syndrome virus (WSSV) in shrimp and prawn. An ELISA-based kit has been developed for detection of IgM of Indian major carps against Aeromonas hydrophila infection. An anti-WSSV formulation was made as a prophylactic measure in shrimp health management. Biosurfactants production was standardized from marine wastes employing the acinetobacter strains.

New source of molecular fluorescent dyes has been attempted from sea cucumber tissues for bioactive compounds and use in biomedical applications. Bioactive properties of ink gland of cuttlefish have been studied for antibacterial, antiviral and anticancer agents. Cell culture attempted on brackish water species, seabass, could lead to establish stable species-specific cell lines useful in virus isolation. The growth-promoting factor studied on development of embryos of Indian horseshoe crab showed significant enlargement of organ development. Development of culture technology in non-conventional species like seaweeds and product and process development of commercial importance was also the part of the programme. A protocol on better aquaculture management practices was developed through recirculatory and raceway system. In service training programme in molecular biology for fisheries scientists was received greater attention in the programme funded at CIFT, CIFA, CIFE. Two programme supports on Aquaculture/Marine Biotechnology have been supported.

DIAGNOSTICS AND VACCINES DEVELOPMENT

At College of Fisheries, Mangalore, work was carried out on the cloning and expression of virulence genes of Aeromonas hydrophila, as an important fish pathogen and the attempts were made to isolate the proteins and study the immunogenicity. At Centre for Biotechnology, Jawaharlal Nehru University, New Delhi, recombinant porin has been expressed and purified in large quantity for maintaining the consistency of the porin vaccine.

The efficacy study and challenge experiments are being undertaken on immunization of mice and fish with porin DNA vaccine. Studies were undertaken on pathogenecity of the Epizootic Ulcerative Syndrome (EUS) in fish at Viswa Bharti University, Santiniketan; North Bengal University, Darjeeling and Centre for Cellular and Molecular Biology (CCMB), Hyderabad.

A project was undertaken at College of Fisheries, Mangalore on expansion of panel of monoclonal antibodies and standardization of immunodot test kit for detection of white spot virus of shrimp. A simple monoclonal antibody (MAb)-based Immunodot kit for detection of WSV of shrimp at the farmers level has been developed, which is a cheap and effective alternative for expensive and complex DNA based tests. Development of DNA– and protein-based diagnostic methods of white spot disease of shrimp, their field evaluation and monitoring were undertaken at Anna University.

The purified viral protein was used as diagnostic assay kit and transferred to Posideon Aquatech Limited, Chennai for commercialization. An ELISA test has been developed using rabbit antisera against carp IgM for the detection of IgM in IMC. Studies were undertaken on development of anti-WSSV formulation against anti-white spot syndrome virus compound as a prophylactic measure in shrimp health management at CAH Abdul Hakeem College, Melvisharam. A process to purify the compound responsible for inactivating the WSSV was optimized.

CELL CULTURE IN AQUACULTURE

Development and characterization of cell lines from selected finfish and shell fish was undertaken at CIFE, Mumbai and a cell culture system was developed from the fish Lates calcarifer (Sea bass). At CUSAT, the cell lines developed from eyestalk sub-cultured and made available to develop such cell lines as and when required.

The cells shown differentiation in to a proliferating bunch of cells and subsequently got detached as a lump of cell aggregates. The development of cell lines from seabass, Lates calcarifer was attempted at CAH Abdul Hakeem College, Melvisharam and CUSAT. The kidney cells shown fibroblastic and formed a monolayer. The cells were were deposited at NCCS.

Growth Promoting Factor(s)/Signal Transduction

A collaborative project was undertaken on identification, isolation and characterization of growth promoting and differentiation factor(s) from perivitelline fluid of developing embryos of the Indian horseshoe crab at Agharkar Research Institute, Pune, National Centre for Cell Science (NCCS), Pune and National Institute of Oceanography (NIO), Goa. Active fraction identified shown significant enlargement and enhanced compartmentalization of the heart in cardiac development and promoting activity. In another project on large-scale cultivation of amoebocytes in vitro from Indian horseshoe crab implemented by NCCS and NIO, regeneration of gill lamellae of the Indian horseshoe crab was successful in amoebocyte production. A molecular signal intervention mechanism was developed as an alternative to eyestalk ablation at the Departments of Zoology and Biochemistry, University of Madras. A specific inhibitor that intervenes in this pathway has been shown to reduce the molt cycle of the prawn and reduces the turnover time and rearing cost and maximizes the economic returns. By decreasing the molt cycle duration, it can accelerate the growth rate of the prawn resulting in early maturation.

Studies on feed supplementation of highly unsaturated fatty acids (HUFAs) and immunostimulants for disease resistance in freshwater prawn was carried out at the Department of Aquaculture, College of Fisheries, Mangalore. HUFAs namely Eicosapentaenoic acid (C20: 5n-3) and Docosahexaenoic acid (C22: 6n-3) are important for larval growth. The supplementation of PUFA incorporated feed could enhance growth, survival and disease resistance in freshwater prawn. A project on prototyping of raceway based third generation shrimp production technology was implemented at the Fisheries Biotechnology Centre, Fisheries College and Research Institute, Tuticorin. The protocol has been developed to increase survival of post-larvae. This in-door shrimp farming facility will be eco-friendly, economically viable and technically useful for large-scale adoption in aquafarming systems in India.

The work on development of Artemia franciscana culture in the solar salt ponds and processing of Artemia cyst and biomass was undertaken at Centre for Marine Science and Technology, Manonmaniam Sundaranar University, Rajakamangalam, Tamil Nadu. Artemia production was initiated in condenser ponds and inoculated in culture ponds and the efforts are being made to utilize the abandoned area of the saltworks for Artemia cultivation. Studies were undertaken at NIO, Goa to isolate, purify & characterize new natural fluorescent dyes and fluorescent molecules from tissues of sea cucumber Holothuria scabra and to test them as bioactivities compounds for therapeutic and industrial applications.

At Kerala Agriculture University, Kochi, attempts were made to isolate bioactive substances such as antibacterial, antiviral and anticancer agents from the accessory nidamental gland and ink of cuttlefish (Sepia pharaonis ehrenberg). An antitumour peptidoglycan was isolated from cuttlefish ink and the crude preparation showed antitumour activity against mice following intraperitonial administration. The peptidoglycan was made up of protein and uronic acid rich with polysaccharide fraction and was found to be made up of five amino acids. The screening of marine Acinetobacter genospecies for production of biosurfactant/s, purification and antimicrobial activity was undertaken at Pune University. Untapped natural sources were screened for novel biosurfactant/s and bioemulsifier-producing microorganisms using Acinetobacter strains, which showed emulsification of toluene, diesel, hexadecane and crude oil, haemolysis of human RBC's and anionic biosurfactant/s production.

At the department of biotechnology - CUSAT, the work on molecular cloning was undertaken to develop a technology for production of alkaline protease enzyme using marine fungus, Engyodontium alba. The optimal conditions were studied for enzyme production using solid-state fermentation of wheat bran as the solid substrate. The enzyme was found relatively stable for one year. At NIO, Goa, fungi from deep-sea sediments were explored as new sources for protease. Fungal growth and protease production at low temperature and elevated hydrostatic pressure was studied to understand their adaptations to extremophilic conditions.

SEAWEED CULTIVATION

At CSMCRI, Bhavnagar, the project was undertaken on scale-up cultivation and processing of phycocolloid seaweeds (Gelidiella acerosa and Eucheuma). Multilocational trials undertaken to validate different culture methods for large-scale cultivation of Eucheuma proved that the raft cultivation method consistently yields better harvest. This method has been successfully demonstrated in the Gulf of Mannar, Tamil Nadu coast as well as at Okha, Gujarat coast. High quality agar was extracted from Gelidiella acerosa and Gracilaria sp. The Gracilaria agar had very high gel strength (1200 g cm-2) with low gelling temperature. This agar will be useful in bacteriological and molecular biology applications.

The nitrifying bioreactor technology developed at CUSAT paves the way for transforming the open prawn/shrimp larval rearing systems to closed re-circulation systems. The reactor, on properly activating and deploying in the hatchery system, can take care of the entire water quality problems. Apart from nitrification this duel capability lowers down the biological oxygen demand, total bacterial population along with stabilization of Vibrio population. The training programme for fisheries scientists was undertaken at three institutes namely Central Institute of Freshwater Aquaculture (CIFA), Bhubaneshwar, Central Institute Fisheries Technology (CIFT), Cochin and Central Institute of Fisheries Education (CIFE), Mumbai to train in service fisheries scientists on aspects of modern biology, health management and conservation of resources etc. Trainees are being given both theoretical hands-on training and independent project work on molecular biology aspects relevant to fisheries topics. After successful completion of the training programme, the trainees have started working in the area of molecular biology in marine sciences at their institutes.

New Programme Support

The programme on aquaculture and marine biotechnology has been identified for support after a detailed discussion under the Task Force on

Aquaculture and Marine Biotechnology. Two proposals from the UAS, College of Fisheries, Mangalore and the Cochin University of Science and Technology (CUSAT), Cochin have been supported for attention on brood stock and larval health management of black tiger shrimp and freshwater prawn and screening for potential anticancer agents from marine Cnidarians, bacteria & fungi. Genotypic characterization and sequencing of genes of interest of marine bacteria, fish and shellfish immunology and screening for useful natural products from microorganisms, capacity building through hands-on training programmes for scholars and technicians on molecular techniques for detection of pathogens and health management in shrimp hatchery system, microbial taxonomy, crustacean immunology and screening and characterization of marine microbes for useful products etc. are also the part of the programme.

Shrimp Genomics

Shrimp (Penaeus monodon) is a very important species for the country from the commercial viewpoint. Unfortunately, the species is not domesticated and the harvests are mainly from the wild. The reproductive biology is complex and the disease burden very high, especially viral diseases. There are populations, which are naturally resistant to viral attack and could be tapped to our advantage. The network programme on shrimp genomics has been developed covering the aspects *viz.* functional genomics in disease resistance of Penaeus monodon, gene regulation of maturation and breeding in P. monodon, microsatellites and physical mapping of P. monodon and genomics of Penaeus indicus for comparative genomics.

TECHNOLOGIES/PRODUCTS

(a) Development of a heat killed whole cell vibrio vaccine using a virulent strain of Vibrio harveyi was developed at College of Fisheries, Mangalore. This has shown immune response in shrimp (Penaeus monodon). The technology has been transferred to Mangalore Biotech Lab.

(b) A polyclonal antibody based immunodiagnostic kits for detection of different bacterial pathogens in finfish and shellfish has been developed by CIFE, Mumbai. The technology has been transferred to GlaxoSmithKline Pharmaceuticals, Mumbai without any modifications for commercialization. The kits are useful as on-farm, rapid diagnostic kits to detect bacterial and viral diseases of finfish and shellfish.

(c) To overcome the problem of viral diseases in shrimp aquaculture and for simultaneous detection of WSSV and MBV, highly sensitive method "Combi kit – WSSV, MBV", a test kit, for specific detection of pathogens is developed by the Department of Microbiology, UAS, College of Fisheries, Mangalore. The test kit can detect all genotypes of WSSV and can be used to detect the presence of virus in shrimp, pond water, sediment, carrier animals etc. The technology has been transferred to Mangalore Biotech Laboratory for packaging and commercialization.

(d) A polyclonal antibody-based immunodiagnostic assay for the detection of white spot syndrome virus - a simple diagnostic test kit developed in joint collaboration with the Centre for Biotechnology, Anna University and C. Abdul Hakeem College, Melvishram was transferred to an aquaculture-based company (Poseidon Biotech, Chennai)

Patent Obtained

In vitro production of amoebocytes from Tachypleus gigas in leibovitz culture medium (U patent # 6,790, 659 dated 14 September 2004) by Ramesh R. Bhonde and Anil Chatterji (2002).

RESEARCH PRIORITIES FOR FUNDING

The support is continuing in the area of Aquaculture and Marine Biotechnology and the emphasis is on new programmes in the following priorities areas identified:

- Structural and functional genomics related to indigenous freshwater/ estuarine/ marine species
- Marine pharmaceuticals; expression of novel genes coding for bioactive macromolecules of therapeutic and industrial applications
- Marine biomaterials, bio-adhesives, bio-flocculent, bio-surfactants, medical implants, biopolymers and bioplastics
- Novel enzymes, biosensors
- Bioremediation/antifouling agents from aquatic organisms
- Development of diagnostics kits and recombinant vaccines for disease management in aquaculture
- Development of cell/tissue culture in cultivable marine organisms, invertebrates and seaweeds etc.
- Fish transgenics for therapeutic and ornamental purpose
- Marine extremophiles.

POTENTIAL AND LIMITATIONS OF APPROPRIATE BIOTECHNOLOGIES

The concept of appropriate biotechnology promoted by FAO refers to the identification of those biotechnological tools which contribute to sustainable development: they are technically feasible, environmentally safe and socio-economically and culturally acceptable. FAO's policy considers biotechnologies as part of an integrated strategy to solve agricultural problems and not as goals in themselves. It must be emphasized that biotechnology transcends the methods hitherto specific to crop or animal improvement and that in both fields it depends on molecular and biochemical techniques to utilize the genetic diversity for improvement of genetic stocks and to confer resistance to major diseases and environmental stress. In some countries, appropriate biotechnologies have been applied successfully in both plants and animals. Agricultural biotechnology cannot be considered independently from the trends in agricultural production.

PLANT BIOTECHNOLOGIES

There are some important aspects to consider in regard to biotechnology implementation in the Near East Region. Although climate conditions make agricultural activities difficult and sometimes unpredictable, the region is one of the world's oldest agricultural areas where biotechnological practices, such as fermentation, began. Therefore, the accumulation of ancient agricultural practices passed down through generations of farmers, along with a rich genetic diversity, provides formidable challenge for modern biotechnology implementation. As current biotechnological techniques have been developed for the needs of advanced countries, the basic biological principles are the same. Therefore countries of the region should likewise adapt or develop biotechnologies for local agricultural conditions.

In order to respond to market demands, and for production purposes, plant breeders have combined their efforts to provide new, uniform and genetically more productive varieties; these, in some cases, can replace the enormous diversity of local heterogeneous or wild varieties which have for centuries responded to agriculture's needs. It should not be forgotten, however, that the plant breeder's starting point is the genetic diversity represented by the heterogeneous materials of the traditional farmer (Villalobos, Ferreira and Mora, 1992).

The extinction of some plants from their centre of origin and their replacement by "improved" cultivated varieties, in particular, has contributed, in some cases, unintentionally, to the decrease in natural genetic variability. Moreover, owing to the growing demand for food, plant breeders have produced very productive cultivars with a very narrow genetic base. In this sense, increasing the yield of a cultivated species may be to the detriment of its adaptation capacity and can increase genetic vulnerability because of the selective pressure during the breeding programme.

For practical implementation of plant biotechnology, access to genetic diversity is essential. Plant germplasm is a limited natural resource that supplies the genes which are essential for the production of improved plant varieties. Genes are dispersed in domestic species and natural populations that have been selected by nature and by farmers over thousands of years according to their adaptation characteristics, resistance and productivity.

Without these resources, the improvement of varieties in farming would not have come about. In recent years, different factors such as the substitution of local varieties by hybrid seeds, excessive application of pesticides, high selection pressure, etc. have caused the extinction of much valuable material that had barely been exploited.

The biotechnology is composed of two major groups of tools which, to a certain extent, are interdependent:

- plant tissue culture; and
- genetic engineering or recombinant DNA.

The following biotechnologies are available for crop production and can be used in the Near East Region:

1. rapid propagation of useful micro-organisms such as nitrogen fixation (NF) bacteria and biocontrol agents;
2. micropropagation of plants, especially when combined with disease-indexing, to produce large quantities of clean planting material;
3. diagnostics based on the use of monoclonal antibodies and nucleic acid probes for the identification of plant diseases and the detection of high levels of chemicals, such as pesticides in food;
4. genetic engineering of individual species to introduce novel traits; and
5. new genetic mapping technologies as an aid to conventional plant breeding programmes.

The major available biotechnologies are being applied successfully to different groups of plants. Each tool is the result of many years of research, trial and error. The justification for the application of a specific biotechnology will depend directly on the critical analysis of the problems to be solved and the availability of suitable techniques. In some cases, conventional methods are more reliable than the most advanced technique.

EFFICIENT UTILIZATION OF FOREST RESOURCES

Biotechnologies provide several choices for use in the efficient utilization of forest resources. However, their main contribution is in saving time in the genetic improvement process, given the long cycle between sexual reproduction in the majority of economically important trees. This situation which is inherent in woody species, makes conventional methods difficult to implement. The major actual and potential applications of biotechnology to forestry are:

- large-scale multiplication of superior trees within the framework of existing dynamic tree improvement programmes;
- genetic improvement for biomass production, resistance to diseases, quality, growing ability, etc.;
- germplasm conservation, particularly of trees with recalcitrant seeds; and

- identification of biochemical markers for germplasm characterization or desirable genetic characteristics.

For forest tree breeding, genetic engineering techniques will allow direct desirable gene manipulation in one step. In the future it will be possible to incorporate genes codified for specific characteristics into commercial species.

Micropropagation of control-pollinated individuals has been accomplished in several woody species, mainly conifers, however, recent contributions have also shown success in other woody species such as Eucalyptus, Leucaena, Hevea, Palms, Prosopis and Casuarina.

Although the list of trees that can be multiplied by tissue culture is constantly increasing, there are still some limitations for broader commercial development of those technologies as alternatives for the use of cuttings in forestry; limitations include the cost of planting stock and insufficient information regarding field performance . Recent developments with somatic embryogenesis systems offer hope on both counts.

Perhaps the best micropropagation system to reach the commercial stage is with radiate pine in New Zealand.

Estimations of the efficiency of this system indicate that one sexual embryo of one of the more responsive genotypes could yield 260 000 plants ready for the field in two and a half years. Other efficient systems are Eucalyptus poplars and Cryptomeria .

Some of the limitations to the extension of these results to other species are related to the lack of response among genotypes of cells cultivated in vitro and reliable mechanisms for many of the commercial species are not available. Other limitation to massive tree multiplication is the adaptation to site in the case of genetically homogeneous material. So far, the use of juvenile tissues as explants is the most commonly used procedure, especially for somatic embryogenesis.

The practical application of tissue culture will be extended to other species, adapting the results of species of proven feasibility and which are used as biological models. However, for full utilization of the potential of biotechnological tools to forestry, the following areas need to be developed further:

- research in haploid plant production;
- studies in symbiosis with mycorrhiza;
- physiological studies in the juvenile-maturation process;
- studies in the regulation of organogenesis and embryogenesis;
- studies in the differentiation of the root system;
- identification of biotechnical markers for early screening;
- studies in morphogenesis of mature tissues;
- development of artificial seed technology; and
- studies in cryopreservation.

IMPORTANT COMPONENT OF THE AGRICULTURAL ECONOMY AND ANIMAL BIOTECHNOLOGIES

Livestock are an important component of the agricultural economy of the region as they provide food, industrial raw materials, traction and manure. There is considerable scope for improving productivity in animals in developing countries. Biotechnology offers new approaches to animal production and health which could benefit the region. Possible applications include: multiple ovulation and embryo transfer; monoclonal antibodies and clones DNA probes; genetically engineered and molecular vaccines; and production of transgenic animals to introduce desirable genetic traits such as disease resistance. Despite the importance of biotechnology in animal production and health, at present there are no such activities in the region. However, FAO has recognized the importance of biotechnology in developing countries and has initiated activities which will eventually include the Near East region.

In October 1986 in Rome, FAO hosted an Expert Consultation on Biotechnology for Livestock Production and Health. This consultation confirmed the need for the application of new biotechnological procedures in order to improve animal health in developing countries. Two main areas were identified for potential implementation: disease diagnosis and new approaches for vaccination and disease control. The participants analysed the potential use of nucleic acid probes for diagnosis, monoclonal antibodies, diagnosis in field conditions, biotechnology in tick-borne diseases, peptide vaccines, baculuvirus and other virus vector vaccines, and the issue of public health implications of biotechnology. Special sessions on regional issues were also held.

In the case of the Near East the expert consultation covered livestock production issues such as: improvement of the nutritive value of straw by biotechnological treatment; manipulation of rumen functions and maximization of livestock production through an integrated system; and recycling, employing the incorporation of by-products (straw) and animal wastes (poultry manure) in animal feeding as well as for the production of energy (methane).

The consultation was followed by two other important meetings concerned with animal biotechnology implementation in developing countries: the FAO Expert Consultation on Applications of Biotechnology in Livestock Production and Health in Developing Countries, held in Havana, Cuba, in 1988, and the FAO/UNDP Workshop on Biotechnology in Animal Production and Health in Asia and Latin America, held in Beijing, China, in 1989. The issues covered are also relevant to the Near East. Other expert consultations held at FAO Headquarters in 1984 and 1992 gave specific recommendations and addressed the problem of newly emerging diagnostic or vaccines production methods based on modern biotechnology. To facilitate the access of developing countries to new biotechnologies in animals, FAO has developed the following major networks:

- FAO Technical Cooperation Network on Animal Production Health Biotechnology. Consisting of different institutions in Latin America (nine countries) and Southeast Asia (seven countries), the network deals with biotechnology related to veterinary diagnosis and vaccines. This network can be extended to the Near East region since it deals with the same problems.
- Computer Assisted Network on Nucleic Acid and Protein Sequencing in Latin America/Caribbean (CA NAPS). This network assists molecular biologists from 12 countries of the region. Since 1989, FAO has been organizing the CANAPS network for developing new vaccines and specific diagnostic tools such as nucleic acid probes. About 100 laboratories have been interlinked in this network and eventually CANAPS will be linked to the International Centre for Genetic Engineering and Biotechnology (ICGEB) in Trieste and (EMBL) in Heidelberg, Germany. Access to this network will be possible for Near Eastern countries.
- Eastern/Central European Network on Biotechnology. Since 1991, the FAO Animal Health Service has been actively involved in the development of a network on veterinary biotechnology related to diagnosis and vaccines covering Hungary, the Czech Republic, Slovakia and Poland. Access to this network will be possible for scientists of the Near East region.
- FAO/IAEA Network on ELISA Sero-Monitoring. The ELISA assay has been used for rinderpest antibodies sero-monitoring within the Pan African Rinderpest Campaign (PARC). The joint FAD/IAEA Division of Nuclear Techniques in Food and Agriculture in Vienna developed the ELISA sero-monitoring kit and has promoted the application of the kit for serological assessment of postvaccinal immunity during the campaign. The Pan-African Network on ELISA sero-monitoring includes over 20 countries. The same division has also developed standardized ELISA kits for brucellosis and diseases of poultry.

Application of monoclonal antibodies for diagnosis of rabies by the WHO Collaborating Laboratory in Tubingen, Germany, has made possible an epidemiological mapping of the rabies strains. At present, the WHO/FAD Programme of Rabies Control in the Near East and southern African is being developed and will be used on a routine basis. The recent session of the Research Group of the Standing Technical Committee European Commission for the Control of Foot and Mouth Disease in Switzerland last year discussed new techniques for the diagnosis of FMD Polymerase chain reaction (PCR) and monoclonal antibodies. The World Reference Laboratory on FMD (IAH-pirbright, United Kingdom and the FAO Reference Laboratory on Rinderpest

"Peste des Petits Ruminants (PPR)" programme provide expertise and reference services on numerous aspects of related biotechnologies.

FAO, in collaboration with leading research centres in the United States, the United Kingdom, Japan and France, has been promoting the development of recombinant vaccines against rinderpest. Successful attempts have also been made to increase the Plowright's vaccine shelf life by a modified freeze-drying scheme and the application of new stabilizers and a selection of thermostable mutants.

The PANVAC centres have been established since 1986 in the National Veterinary Institutes of Dakar, Senegal, and Debre-Zeit, Ethiopia, in order to strengthen vaccine production and quality control in 23 countries. In Africa, 28 veterinary vaccines are produced by 23 manufacturers. One of PANVAC's objectives is the technology transfer to Africa of bacterial, viral and anti-parasitic vaccines. A strict standardized quality control of rinderpest vaccines initiated by PANVAC resulted in a significant improvement in the quality of vaccines applied in the Pan African Rinderpest Eradication Campaign (PARC). PANVAC has also been assisting the Near East to strengthen capabilities in countries for rinderpest vaccines production and quality control and can be a model for the region.

THE FUTURE OF ANIMAL BIOTECHNOLOGY IN THE NEAR EAST

There has not yet been a full assessment of the existing capabilities of countries with regard to animal biotechnology development. Some of the most advanced countries in the region have, however, been making individual arrangements with regards to biotechnology transfer and training. Some of the needs for biotechnology implementation in the region are:

- to promote international coordination in national efforts on biotechnology development;
- to develop and strengthen veterinary biotechnology for assistance to countries in formulating and promoting national programmes;
- to develop a regional programme on animal biotechnology and coordinate activities at national subregional levels;
- to strengthen training in modern biotechnology at advanced levels and develop a training programme in animal biotechnology for young scientists; and
- to promote development of national and international institutions related to veterinary biotechnologies.

Biotechnology applied to agriculture has hitherto been mainly transferred to developing countries through universities with scientific interests and, often with educational objectives. More recently, international centres such as ICARDA, ICRISAT, and ILCA have been actively participating in the transfer of biotechnology to their counterparts, the national programmes. Although some activities in the application of modern biotechnological procedures are taking place in the Near East region, these initiatives are, in most cases, associated with limited resources, lack basic biological information on species used and lack support. This current situation limits the real potential of biotechnology for solving problems in the agricultural sector. For the appropriate application of biotechnological tools in agriculture, some prerequisites are essential, including:

- an inventory of the existing human, physical and genetic resources available to biotechnology;
- an inventory with a priority of the main agricultural problems and potential techniques adapted to country's conditions;
- existing national programmes for genetic improvement;
- financial and political support;
- scientific knowledge and a positive aptitude for multidisciplinary research; and
- efficient communication systems; access to scientific literature, equipment (maintenance) and chemicals.

If any one of these elements is missing, expectations for biotechnological achievements should not be optimistic. Countries of the region must therefore define their policies on the use of biotechnologies.

FAO is following the recommendations made by the participants in the FAO/CTA meeting held in Luxembourg in 1989 when FAO was requested to create a Programme on Plant Biotechnologies for Developing Countries in collaboration with other development agencies.

With reference to the implementation and adoption of appropriate biotechnologies, FAO is in the process of finalizing a programme which will address the needs and aspirations of member countries, in particular less developed countries.

The FAO programme will strengthen the relevant biotechnological capabilities of developing countries by promoting:

- cooperation between advanced laboratories and national programmes in developing countries, and networking;
- activities aimed at fostering research and development;
- communication and information systems;
- training and infrastructure.

Biotechnology activities will rely on:

- FAO staff having diverse technical experience in agriculture and food production and the relevant policies;
- the Organization's new and ongoing projects and programmes aimed at assisting its member countries;
- contracts and collaboration with leading biotechnology laboratories in developed and developing countries; and

- the forum that FAO, as an international organization, provides for the discussion of important technical, ethical, legal, environmental and political issues.

The FAO/AGP Programme on Plant Biotechnology will assist member countries in identifying their requirements and mobilizing resources for institutional and staff development. The programme also intends to promote regional networks in order to make optimize of scarce labour, equipment and other resources, particularly in developing countries.

PROGRAMME OBJECTIVES

The programme objectives are to:

1. Assist in the integration of plant and agricultural biotechnologies into the national plant breeding and germplasm conservation programmes.
2. Facilitate the implementation of appropriate biotechnologies for sustainable agriculture.
3. Design strategic research and development projects for funding sources.
4. Monitor and appraise the impacts of new plant biotechnologies and inform the member countries through publications and meetings.
5. Promote the exchange of information and experience through networking at different levels.
6. Strengthen the biotechnological capabilities of developing countries in priority setting, personnel training in the relevant plant and agricultural biotechnologies, technology transfer and application to seed production, micropropagation followed by large-scale multiplication, germplasm exchange, etc.
7. Provide a forum for debating intellectual property rights and legal and ethical issues.

One of the FAO/AGP programme strategies is to create a flow of information and knowledge between laboratories in developed or developing countries that are at the frontier of biotechnology research. This flow of information as well as plant germplasm is a two-way system, so that strong links between supply and demand are stimulated. One of the major objectives of the FAO/AGP programme on Plant Biotechnology will be to foster these associations. Also, FAO will act as a liaison, promoting, supporting and stimulating four major kinds of activity:

- FAO field projects in plant and agricultural biotechnologies;
- networking;
- case-studies; and
- information/communication.

These activities will be carried out in close collaboration with national governments, other UN agencies such as UNESCO, UNIDO, UNDP, and funding institutions such as the World Bank, regional development banks, the CGIAR systems and other relevant organizations involving private industry and NGOs. Participation will also be sought from the interinstitutional tasks forces involved in the international or regional plans of major projects.

The proposed FAO programme can be implanted in the Near East region with the support of national governments. The strategy could be the implementation of a regional network for technology transfer and training. The proposed network could facilitate the information exchange among laboratories within the region and also from selected advanced laboratories outside the region. Conferences and scientific meetings for information exchange could also be programmed and promoted according to national priorities. Activities on germplasm conservation and use should be approached at the regional level, in which the network could play a leading role. The major goal of this regional programme should be the gradual development and implementation of biotechnology infrastructure, and identification of human resources capable of contributing to solving problems in the agriculture sector of member countries of the Near East region. To initiate actions for the implementation of this programme, the support of the countries of the region will be essential.

BIOTECHNOLOGY FOR THE ENHANCEMENT OF FEED PRODUCTION

Several Asian countries are involved in work on biodegradation of organic wastes for the production of animal feed or biogas. The majority of these studies cannot be categorically called biotechnological work since they involve known and old methods of degrading lignocellulosic materials for improvement of the nutritive value of feed. Therefore, as summarized by Mukherjee (1989), only those projects/laboratory research which involved either microbiological treatment for biodegradation or dealt with huge machineries/fermenters/powerplants are briefly mentioned.

In most of the provinces and autonomous regions of China, microbial treatment of lignocellulosic low-quality roughages is practiced to a varying extent. Yeast factories located at Shanghai, Jilin and Guang Dan, the Nanjing Fermentation factory and the Shunde Sugar factory produce single cell proteins containing 40-85 percent protein. Production of SCP using petroleum hydrocarbon, natural gases and methanol are also commonly practiced. Enzyme product factories in Wuxi and Tianjing produce various types of enzymes. The Shanghai Biochemistry Institute and the Beijing Institute of Biochemistry produce enzymes through genetic engineering work. Furthermore, at least 18 amino acids are produced using fermentation technology. In India, the Indian Institute of Technology, New Delhi, undertook biochemical engineering research for the development of single cell proteins from waste biomass. The National Dairy Research Institute, Karnal, has been researching the enrichment of wheat straw using fungal (Coprinus) and urea treatment.

In Indonesia, the Central Research Institute for Animal Sciences, Bogor, has investigated prospects of protein enrichment of cassava tuber using yeast inoculum (Candida ingenuosa and Candida utalis), protein enrichment of rice straw and wood chips using Pleurotus Protein enrichment of rice straw using Coprinus was studied at the National Centre for Research in Biotechnology.

In Malaysia, utilization of agroindustrial wastes as animal feed using biotechnological methodologies is being emphasized. The role of rumen in the digestion of fibres in ruminants with a view to developing in vitro systems of digestion of fibrous materials has been studied. In this context, characterization of rumen bacteria with a view to manipulation of bacteria in future for effective digestion of fibre materials was undertaken. Production of protein biomass utilizing palm oilmill effluent as a substrate for thermophilic and mesophilic fungi has been explored.

In Pakistan, the National Agricultural Research Centre, Islamabad, has been studying the fungal treatment of agricultural wastes (cereal straws and sugar cane bagasse). The Department of Animal Nutrition, University of Faisalabad, has been researching mycelial biomass protein production from rice husk, rice polish and rice straw using Trichoderma harzanum.

PRODUCTION OF MONOCLONAL ANTIBODIES AND VACCINES

Australia is promoting host specific vaccines. More specifically, viruses that are genetically engineered to infect selectively and prevent reproduction in feral animals are considered a desirable goal achievable in the short to medium term. Other goals include endocrine-directed vaccines to stimulate twinning in beef cattle, immunocastration, livestock growth rate increase, carcass composition such as reduced fat, and vaccines that compensate for stress-induced production losses of various kinds. Diagnostics are being used for quarantine management. Tailored vaccines exist for pig scours, chicken bursar disease and cattle tick. The latter is being promoted particularly in North Queensland where cattle tick causes babesiosis.

Most of the Asian laboratories dealing with the diagnosis of viral and bacterial diseases and production of vaccines have improved their diagnostic techniques during the past few years. The use of immunoflorescence microscope, radioimmunoassay and enzyme-linked immunosorbent (ELISA) techniques are quite common in most laboratories. Among the developing countries, work on the production of monoclonal antibodies has been quite intensive in the Republic of Korea, China, and India. Hybridoma techniques are also being used in some laboratories in Malaysia and the Philippines. Similar work is also in progress in Indonesia, Thailand and Pakistan.

In summary, as regards animal biotechnology in developing Asian countries, while high priority is being assigned to this subject, the need to maintain and further consolidate present work for animal improvement, health and production through conventional means is duly recognized.

Embryo Transfer (ET) has been most actively pursued in all countries. Large-scale public sector funds were available in China, India and Pakistan for the development of ET.

China has developed considerable indigenous technology, *e.g.* production of hormones, embryo freezers and nitrogen tanks, which could be suitable for relatively poorer countries in Asia.

The Republic of Korea's efforts were very practical and included a cooperative approach of government, research institutions, universities and veterinary clinics. Initial private initiatives to develop ET in the Philippines, Thailand and Indonesia have been successful but not sustained.

Genetic engineering research in relation to domestic animals has had a good start in China, the Republic of Korea, India and Malaysia where substantial commitment has been made by the R&D funds of the respective governments. There are capable scientists in several Asian countries who could undertake molecular biology work, but because of limited research funds, such work has not yet gained momentum in many institutions.

Recognizing that the inadequate availability of animal feed is a major bottleneck in the development of livestock industry in the region, Asian countries are keen to promote the use of biotechnology in converting agricultural by-products and wastes into animal feed. Microbial digestion of raw materials using fungi and bacteria is being widely pursued. Characterization of rumen microbes in relation to their properties of digesting various fibrous feed materials will continue to be an important research subject in animal biotechnology in the region.

The discovery of monoclonal antibodies for the diagnosis of different diseases in the Republic of Korea, India and a few other Asian countries has stimulated industrial activities and the trend must spread to other countries in the region.

RECOMBINANT DNA TECHNOLOGY AND ITS APPLICATIONS

Recombinant DNA technology research in Asia is mainly confined to basic research by molecular biologists involving prokaryotes. This entails the discovery of suitable vectors (plasmids and nonplasmids) for the propagation of genes and the expression of these genes in selected hosts. Some biological laboratories are also engaged in DNA fingerprinting or restriction fragment length polymorphism work. Most of these biological laboratories are adequately equipped to undertake major research and train scientists from agricultural disciplines.

Only a few animal science/veterinary laboratories in the developing countries were engaged in molecular biology research. There are very few published reports on the work done, hence the information given below is based on information contained in various reports of the regional FAO/UNDP project on animal biotechnology mentioned earlier. In China most of the work is being conducted by the National Laboratory of Agro-biotechnology, Beijing Agriculture University. The following are the highlights of the work in hand:

- The specific gene of equine infectious anemia has been successfully cloned.
- RFLP is used in recharacterizing of some indigenous animal breeds and in classifying some wild animals.
- The gene mapping method is used to identify resistant genes in the MHC B-L subregion of O-type chickens, and to identify other related genes in the vicinity of GH genes in pigs.
- The following gene constructs have been made: bovine genomic growth hormone gene, and procine genomic growth hormone gene, sheep genomic growth hormone gene and procine growth hormone releasing factor gene. Genes for production related hormones (calf renin, bovine pepsin and buffalo renin) have been constructed.

In India major work has been undertaken since the 1980s by the Indian Veterinary Research Institute (IVRI), Izatnagar and Bangalore campus and the National Institute of Immunology (NII), New Delhi, in the following areas.

- Molecular characterization of pox virus, especially caprine virus (both virulent and non-virulent types) is in progress using gene sequencing techniques. Identification of promoter region in the genome is being made in collaboration with the Virus Research Institute, Pirbright, United Kingdom.
- Studies on molecular cloning in Clostridial species for vaccine production have been initiated. Early work involved DNA isolation of non-pathogenic Clostridial species (C. perfringenes, type C and type D), its characterization and isolation of plasmid DNA from host vector.
- Another group of workers has been engaged in the characterization of gene sequences in Clostridium welchi type D and location of gene blocks, responsible for the production of "alfa toxin".
- Molecular characterization of FMD virus type, O, A and C has been attempted. Complementary DNAs for viral RNA type A and O were prepared. A c-DNA probe has been successfully used for identification of virus specific sequences.
- Cloned-DNA for the major antigen of FMD disease has been shown to be expressed in E. coli.
- A partial library of Y chromosome derived DNA sequence of bovine origin in E. cold has been constructed at NII.

- Characterization of extra chromosomal plasmids responsible for desired characters of starter culture and identification of plasmid for lactose utilization in strains of streptococci.

In Japan, research in transgenic animals is highly advanced. In 1989, the National Institute of Animal Industry, following work in the United Kingdom, succeeded in in vitro culture of poultry eggs, achieving a good hatchability of about 34 percent. These results greatly increased the feasibility of transgenic chickens. As regards animal health, a recombinant vaccine of bovine Iymphatic leukaemia virus has been developed by the National Institute. A rabies virus vaccine was also developed. Highly specific and sensitive antibodies have been designed and are being used for the diagnosis of bovine Iymphatic leukaemia, rotavirus infection, mycobacteriosis and paratuberculosis, which are difficult to detect by traditional diagnostic methods .

Japan is harvesting biotechnology for altering silkworms to increase the production and quality of silk filaments and to find new horizons of use other than traditional silk production. Using recombinant DNA technology, the National Institute of Sericultural and Insect Agricultural Technology is transferring the unique glossy yellow greenish silk colour from wild silkworms to domesticated silkworms. A technique of utilizing silkworms as the "factory" for producing useful substances has been established. In this method, for instance, the genes interferon are incorporated into nuclear polyhedrosis virus, and this virus is propagated in the body of silkworms so as to be mass produced.

This technique was found to be more efficient than using bacteria as hosts and will be applied to the production of many other useful substances in the years ahead . The structure and function of hormones controlling the growth and metamorphoses of insects have been understood, and as a result it is now possible to control the growth of insects to some extent. Using this technique to control Silkworm larval cycle and maturity, it is now possible for Japanese scientists to control the thickness of silk filaments from one-tenth to three times that of ordinary ones .

In Malaysia, genetic engineering research for animal production and health has been undertaken by the faculty of Biotechnology and Food Sciences and the Veterinary Faculty of University Pertanian Malaysia (UPM) and two Departments of University Malaysia (UM)-the Department of Genetics and Cellular Biology and the Institute for Advanced Studies. Research in progress or completed include:

- Molecular characterization of Newcastle Disease (NCD) virus.
- Collection of baseline data on diversity of rumen bacteria and fungi has been completed and studies on genomic and plasmid DNA in selected microbes and fungi have begun.
- Cloning of xylanase gene for bacteroides succinogens, using E. cold HB 101 as the host.

- Mitochondrial and nuclear RFLP work as markers for the identification of different breeds of buffaloes and chickens is in progress. The cloning of pituitary growth hormone gene using retroviruses as a vector was initiated.
- A study has been completed to relate the transformable antibiotic resistance traits in E. cold strains of bovine, avian and procine sources from slaughterhouses with the frequency of R plasmids, which are extrachromosomal genetic markers, and are suspected to be responsible for drug resistance.

In the Philippines, basic work on several aspects of animal health biotechnology is being conducted at the National Institute of Applied Microbiology and Biotechnology, Los Baños. The Institute uses protoplast fusion and recombinant DNA technology in the genetic improvement of antibiotic producing microbial strains. Research on local production of tylosin and antibiotics for animal feed is also in progress.

The National Biotechnology Institute, Los Baños, has also undertaken work on the development of low cost biogas systems using crop residues, production of genetically improved lignolytic micro organisms, enzyme engineering techniques for synthesis of ligniases, and the low technology tumbler process of producing microbial proteins using selected strains of fungi and yeast which could be used as a feed ingredient for poultry and pigs.

As regards other Asian countries, the Veterinary Faculty of Kasetsart University, Thailand, the Central Research Institute for Animal Sciences, Bogor, Indonesia, and the Department of Animal Sciences, Agriculture University at Faisalabad, Pakistan, have initiated work on the identification of genetic markers for production and disease traits in livestock.

COMMERCIALIZATION AND ROLE OF THE PRIVATE SECTOR

In general, commercialization of agricultural biotechnology is far behind the application of biotechnology in human health and medicine. In the developed countries, especially in Japan, the private sector plays a major role in the generation, development and commercialization of biotechnology products and techniques. Because of commercial considerations, concentration has been mainly on diagnostics, vaccines, pharmaceuticals and other health-related products. Moreover, because of the eminent involvement of the private sector, marketability of the products and potential return on investment are crucial factors in deciding which products are to be developed and commercialized.

In Japan, the MAFF is subsidizing fundamental technical development projects in the private sector and providing funds for investment in such projects. It is also offering financial assistance to prefectural governments so as to encourage their R&D activities and utilization of the outcome of such activities. The Ministry takes steps to subsidize the research projects of private technical research associations and other similar study programmes. These subsidies are offered mainly through the Ministry's project for the development of biotechnology and other high technology for the food industry. In fiscal 1989, MAFF began such financial assistance programmes as the "Transformation of plant cells through the introduction of small organelles into the cells" and "Genetically manipulated vaccines against animal protozoa diseases" .

In the area of biotechnological R&D activities in the food industry, the Ministry started the project "Development of bioreactor systems for the food industry" in 1984. This project has achieved a number of remarkable results regarding the conversion of saccharides, fat and other substances. Another project, the "Development of technology for improvement of enzyme functions for the food industry" aims to improve the functions of enzymes for food production utilizing recombinant DNA technology. In fiscal 1989, the Ministry began a new programme, "Food production by large-scale, high-density fermentation under ultra-high pressure conditions" with a view to increasing the efficiency of the food production process under ultra-high pressures and to finding food materials with a higher value added.

Two more projects were launched in 1989: the "Utilization of extracts from trees (green sprouts project)" and "Advanced and multipurpose utilization of unused marine resources (marine-frontier project)". The objective of these projects is to develop the technique for extracting useful substances from trees and marine organisms and utilizing them for foods, perfumes and medicines, thereby helping further growth of forestry and fishery industries.

The Bio-Oriented Technology Research Advancement Institution (BRAIN) was founded in October 1986 by the joint investment of the government and the private sector. In an effort to encourage the R&D activities of the industries related to agriculture, forestry, fishery and food production, this institution is investing in joint technical development corporations and is providing no interest-bearing loans to the technical research projects of businesses on certain conditions.

During three years from fiscal 1986 to 1988, BRAIN invested in 14 study projects and offered loans to 70 projects. In fiscal 1989, the Institution invested in, among others, the Institute of Aquacultural Technology of Cold Water Fishes, Wakayama Agro-biological Research Centre and the Japan Turf Grass Co. The first is engaged in the development of aquaculture systems of cold water high-class fishes, the second in R&D activities for rationalized process of tangerine juice production and the third in the breeding of better turfs.

In the area of biotechnology-related R&D in the private sector, financing systems through the Japan Development Bank, the Agriculture, Forestry and Fishery Finance Corporation and other similar government organizations were reinforced to turn the results of these R&D activities to practical use. It is hoped that as biotechnology is increasingly introduced to industries, these financial measures will be utilized more in the coming years.

Prefectures are becoming more and more interested in R&D activities on biotechnology and the practical use of the outcome of such activities. This is evident from the fact that almost all prefectures throughout Japan have a council on this technology composed of scholars and other people of expertise.

MAFF is offering to prefectures a variety of subsidies in order to diffuse the mass production technology of virus-free seedlings of vegetables, ornamental plants and fruit trees and the transplanting technology of the embryos of beef cattle. In fiscal 1989, the Ministry started the "Project for establishing embryo supply centres" to ensure a stable supply of embryos of beef cattle. This project is improving the raising and controlling facilities of cattle from which such embryos are taken.

Since 1986, MAFF has also been implementing the "Project for assisting local R&D activities in biotechnology". This project offers financial assistance to the R&D activities of prefectural research institutes concerning the breeding and utilization technology of biological resources. Other biotechnology-related activities of the Ministry include regional biotechnology meetings held in seven locations across the country with the object of exchanging information on R&D. The Ministry's new initiative in fiscal 1989 included surveys and studies for examining the direction of the practical use of biotechnology in each region. The Tsukuba Bioscience Hall created in 1989 provides a forum for collaboration among trainees from industries, universities and the government and international exchange programmes. Furthermore, to foster links among genetic resources and biotechnology, the Ministry has a strong national genebank and germplasm conservation programme .

Of the developed countries in the region, the involvement of the private sector in biotechnology, especially plant biotechnology, is rather limited in New Zealand. Very little research on plant biotechnology is being carried out by private companies. A shortage of venture capital, low taxation incentives and tough competition from large overseas firms make such ventures unattractive.

Several small companies are engaged in either product formulation, or manufacture of biological control agents. Larger international companies (*e.g.* Monsanto, MPI Koln, PGS Ghent, ICI) and overseas research institutions (UCLA, California; John Innes Institute, United Kingdom; Christian Albrecht University, Germany) have provided personnel, facilities and some supportive funding on collaborative projects. Producer Boards (*e.g.* New Zealand Kiwifruit Marketing Board) also support research through joint funding of government programmes .

Many New Zealand research programmes on plant biotechnology are not mature enough to have reached the stage of successful commercialization. In addition, some programmes are commercially sensitive, particularly where research is collaborative with private companies. Examples of plant biotechnology research that have been, or are in the process of being commercialized in New Zealand include :

- Establishment of a commercial tissue culture laboratory for clonal Forestry with Radiata Pine (FRI).
- Use of immuno-molecular techniques for developing test assay kits and for analysis of gene expression (DSIR, Fruit and Trees).
- Plant variety rights have been granted for four new plant varieties.

Future commercialization is likely to be through the release of improved crop varieties, development of new cultivars and the licensing of technologies that have been developed.

In the developing countries, in vitro culture for micropropagation of elite and disease-free materials, production and distribution of efficient nitrogen-fixing microbes, diagnostic kits and monoclonal antibody-based vaccines are the main areas that have been commercialized to varying degrees in different countries.

In developing countries, the role of the private sector in modern biotechnology has been rather limited. This is attributed partly to: (i) vague government policies regarding the private sector; (ii) unclear policies or no policies/views on patents and intellectual property rights; (iii) poor links between public and private sectors; (iv) low purchasing power for new products that are usually highly priced and are out of reach of the majority of resource-poor farmers and low-income consumers; and (v) inadequate local expertise and infrastructure and R&D support to new biotechnology.

In India, biotechnology industries have recently been growing in the private sector. Hindustan Lever Ltd, an Indian subsidiary of the multinational Unilever, the Tata Energy Research Institute (TERI) and Southern Petrochemical Industries are some of the leading biotechnology companies in the country. These have considerable in-house R&D bases active in the application of new technology to agribusiness, such as enzyme technology for high-value chemicals and quality edible oil by using genetically modified bacteria; protoplast fusion in yeast for efficient fermentation and production of edible oil from biomass; tissue culture of plantation and ornamental crops; biological nitrogen fixation; organic components for higher rate of photosynthesis; big-insecticides; plant growth promoters; biomass processing for animal feeds; and hybrid seeds. Some small and medium-size companies are also initiating biotechnology activities but, with less R&D capacity, are more wary of what they consider to be a higher-risk technology.

The private sector in India is particularly active in tissue culture products. There are about 50 private sector companies such as Indo-American Hybrid Seeds, A.V. Thomas and Co., Unicorn Biotek, the three companies and others that are comprehensively involved in commercial in vitro production of planting materials of several horticultural and plantation crops such as cardamom, coffee, tea, oilpalm, orchids, strawberry, roses, Spatiphyllum, banana, lily, Gerbera, etc. both for domestic and export markets. These companies have established effective links with the public sector .

The recent liberalization of trade, including that of certified seed, coupled with the greater priority given to science and technology and the promotion of private sector and public-private sector links, should encourage increased participation of the industry in agricultural biotechnology R&D. Financial incentives are available for supporting the growth of indigenous biotech industries. However, venture capital is still lacking. Several multinationals, *e.g.* Seedtec International, Cargil Inc., Dehlgien Inc., Northrup King and Sandoz (through their local subsidiaries) have been shortlisted by the Indian Government for production of hybrid seeds which may involve biotechnologically produced seeds in collaboration with Indian companies. To take advantage of the availability of expert personnel in biotechnology at relatively low costs, a few multinational companies like Astra AB (Sweden) are setting up a biotechnology R&D centre in India. Recently, Pro-Agro, an Indian seed company, established links with Plant Genetic Systems, a leading Belgium-based plant biotechnology company, for the production and distribution of hybrid seeds.

In Indonesia, there is relatively little direct involvement or sponsorship of research in government institutions at present. Some companies are investigating improved planting materials. A commercial production system has been started for improved oil palm plantlets, with a target of 1 million plants per year in 1993, at the Research Institute for Oil Palm at Marihat. Animal vaccines are produced by Pus Vetima and human vaccines by Perum BioFarma, in government-owned companies, which also undertake some biotechnology research.

In the Republic of Korea, the entrepreneurship drive led the biotechnology sector to be organized in 1982 under the Korean Genetic Engineering Research Association (KOGERA). At present more than 20 private companies are participating in this Association. The government provides the members of KOGERA with subsidy as the seed money. In 1980, the public sector provided 70 percent of the R&D support to the biotech industry. Six years later, the situation was reversed; in 1986 the public sector provided only 28 percent of the R&D cost whereas the industry provided 72 percent of the cost. This shows the increasing success and confidence of the private sector. The private companies are mostly involved in fermentation and pharmaceutical matters and have very little interest in agricultural biotechnology. Most of the companies are using biotechnology for the production of antibiotics, insulin, new vaccines and interferons.

The Association performs an important role in the exchange of biotechnology information and also promotes relevant research. MOST provided a special fund for 122 projects in the field of gene manipulation and recombinant DNA technology in 1990. Of these national projects, 32 projects were directly connected with crop improvement . The Korean Science Foundation (KSF) has contracted research projects to universities relating to biotechnology in the fields of medical, pharmaceutical, and agricultural sciences.

The entire biotechnology work in agriculture in the Republic of Korea is coordinated by the Agricultural Biotechnology Research Council comprising researchers from national institutes and universities and representatives of industries and the private sector. A few private companies involved in agricultural biotechnology are using the new technology for the production of virus-free potato planting material, and through the anther culture technique are producing haploids for heterosis breeding in Chinese cabbage and rice. The private sector, besides having strengthened its own R&D system, also provides assistance to universities for undertaking specific biotechnological research. The increase in personnel and funds for biotechnological research and development in the private sector in the Republic of Korea.

INTELLECTUAL PROPERTY RIGHTS

Two Intellectual Property Rights (IPR) mechanisms namely, patents and plant breeders' rights are most important for agricultural biotechnology. A patent is a right granted by the government to inventors to exclude others from imitating, manufacturing, using or selling a specific invention for commercial use during a certain period, usually 17-20 years.

The patent holder, in turn, is obliged to disclose the invention to the public. Plant Breeders' Rights (PBR) are rights granted by the government to plant breeders to exclude others from producing or commercializing material of a specific plant variety for a period of about 15 to 20 years.

Both under IPR and PBR, use of the protected matter is restrictive, being highly restrictive under IPR. There is, however, provision for research exemption under IPR which allows others to study the protected subject-matter without reproducing or multiplying it for commercial purposes.

Under PBR, there are provisions for breeders' exemption and farmers' privilege. But, in 1991, UPOV tightened PBR by eliminating the breeders' exemption for an essentially derived variety-a variety predominantly derived from another (initial) variety which retains the expression of the essential characteristics from the genotype of the initial variety.

Now, as per the 1991 legislation, a breeder who inserts a single new disease-resistance gene into a PBR-protected variety will have to obtain permission from the holder of the original rights before marketing the new variety. In 1991 UPOV also attempted to tighten up farmers' privilege, *i.e.* freedom to farmers' freedom to use their own harvested material of protected varieties for the next crop on their farm, but because of a lack of consensus among UPOV members it was left to the national governments whether to permit farmers to reuse the seed of a BPR-protected variety for further crop cycles on their own holdings.

The status of development and adoption of IPR and PBR varies widely from country to country in the region. The three developed countries (Australia, Japan and New Zealand) are all members of UPOV and have explicit IPR provisions for biotechnological products and procedures. None of the developing countries are members of UPOV and have mostly been opposed to enforcement of any form of IPR, particularly patents.

However, lately, under various kinds of bilateral pressures, such as the special 301 Provisions of the US Omnibus Trade Act of 1988, as well as multilateral pressures, and after having signed the successfully negotiated GATT agreements in December 1993, whether to introduce legislation covering proprietary protection for biotechnology products and procedures, and/or how rigid the protection system should be, are under active debate in several developing countries of the region.

Those in favour of having the legislation argue that an effective IPR system would: (i) improve the access of the country to new technology from abroad; (ii) stimulate private sectors; and (iii) encourage foreign investment.

Those against it argue that in the context of the developing countries it will: (i) suppress in-country innovation and local industries and production systems; (ii) increase the dependence on foreign countries even for materials and procedures that are vital for the people of the importing country; (iii) increase prices of even commonly used products and services; and (iv) restrict information flow and knowledge growth.

The effects of IPR on the development of biotechnology in developing countries are not clearly understood and should be studied systematically, covering countries at various levels of development and of varying sizes and in the context of different industries, such as pharmaceuticals, agriculture, food industries, environmental protection, etc.

Thailand adopted a new patent act (1992) that protects biotechnology inventions after the United States Government had withdrawn Thailand's trade benefits under the General System of Preferences. Under a similar pressure, when the renewal of the Sino-American Science and Technology Agreement was postponed by the United States Government, China revised its patent law effective as of January 1993, which protects biotechnology inventions, but excludes plant and animal varieties.

In India, the matter is hotly debated. Despite the pressure from the United States Government by postponing the Indo-US Vaccine Action Programme and the Indo-US Science and Technology Initiative, the Indian Government has not yet strengthened its IPR legislation. However, given these pressures, the changed and liberated seed policy of the government and increasing pressure from the private seed companies within the country, there is a change in the mood of the government, although strong opposition persists. India signed the GATT agreement in December 1993. Its 1994 Spring Session of the Parliament held a special debate on the issue, and the opposition parties continue to agitate. It appears that a consensus may emerge to evolve an IPR system that will not promote inequity, jobless growth, and ecological degradation. The government is currently preparing an effective sui generis UPOV PBR, explicitly emphasizing breeders' exemption and a strong farmers' privilege. In addition, the system would provide for farmers' rights and include mechanisms for the operation of these rights.

As regards other developing Asian countries with considerable programmes in agricultural biotechnology, no patent protection systems exist either in Malaysia or the Philippines. However, Indonesia is preparing a new IPR legislation.

It can be seen from the above that, under various kinds of pressures, from outside or within, and as biotechnological R&D efforts are being intensified, several developing countries are moving towards restructuring or strengthening their IPR systems.

What is important at this stage is that the countries strike a rational balance between the positive and negative aspects of the IPR system and tailor their legislations to their needs, opportunities and aspirations. In countries where the local technology base is weak, an IPR system may contribute little to the indigenous technology capacity.

However, if the country is now able to pursue cutting-edge biotechnology and produce innovative and useful products and processes at competitive prices, the relevance of an appropriate IPR would increases. The countries must therefore be able to analyse the various pros and cons and short-and long-term socio-economic and ecoenvironmental advantages and disadvantages of instituting a particular IPR system. International organizations, such as FAO, should assist the countries in this task.

REGULATORY ASPECTS

Biosafety

R&D biotechnology is heavily influenced by two regulatory issues, namely biosafety and intellectual property rights. Biosafety aspects include the policies and procedures needed to ensure the environmentally safe development and application of biotechnologies. An effective biosafety system to regulate release and use of Genetically Modified Organisms (GMOs) and to address concerns about potential risks to public health and environment must be in place in all countries involved in biotechnology. Biosafety regulations are necessary also to facilitate technology transfer. Comprehensive biosafety guidelines and measures exist in the developed countries, but the situation is far from satisfactory in most developing countries, although they are aware of the need and are in the process of formulating national frameworks, procedures and capabilities for addressing biosafety issues.

In Australia, biosafety guidelines/regulations have been evolving since 1975, when the Academy of Science Committee on Recombinant DNA (ASCORD) issued its first guidelines. In 1981, as per the Commonwealth resolution, the Recombinant DNA Monitoring Committee (RDMC) was established. In 1987, the RDMC published "Procedures for the Assessment of the Planned Release of Recombinant DNA Organisms". The RDMC was further transformed into the Genetic Manipulation Advisory Committee (GMAC) to include the surveillance of any genetic manipulation procedure resulting in an organism containing foreign DNA that was unlikely to be formed by conventional breeding practices. The GMAC emphasizes that the focus of the assessment should be directed towards ensuring that the release does not cause harm to people, livestock, plants or the natural environment, and not primarily be directed towards considering how the organism was constructed.

GMAC adopted the following procedure :

1. Approval of the concerned Institutional Biosafety Committee (IBC) is obtained.
2. The IBC sends the proposal to GMAC.
3. GMAC, according to its Procedures for the Assignment of the Planned Release of Recombinant DNA Organisms, in consultation with its Scientific and Planned Release Committees, analyses and evaluates the proposal, especially the genetic aspects and any possible environmental impact associated with the construct.
4. The above assessment report, especially pointing out hazards, if any, is sent to the proposer and to the agency that has legal jurisdiction over the release of that particular class of organism.
5. The responsible agency then makes the decision whether to release or not to release the organism based on the report from GMAC and other related considerations."

In Japan, five guidelines have been established for the regulation of research of DNA organisms and the safe promotion of experiments since 1979. Appropriate regulations are considered effective in minimizing the potential risks of a new technology to the environment and human health, and in ensuring positive public perception. The purpose of MAFF's "Guidelines for the Application of Recombinant DNA Organisms in Agriculture, Forestry,

Fisheries and the Food Industry and Other Related Industries", is to promote the safe progress of agro-industries by: (i) defining general principles for the appropriate application of rDNA organisms; and (ii) ensuring safety in the use of rDNA organisms . The guidelines are based on OECD guidelines.

Applications for field trials of rDNA organisms are to be approved by MAFF through biosafety assessments conducted by the authority committee under MAFF. The first application for field tests using transgenic tomato plants harbouring TMV coat protein genes was filed in late 1990. The application was based on three years of greenhouse experimentation. As the biosafety review of this application was completed, the first field trials were approved in 1991 and since then many field tests have been approved.

In New Zealand, an Advisory Committee on Novel Genetic Techniques (ACNGT) was set up in 1978 under the terms of a Cabinet resolution. All persons working in the public sector (universities, government departments, etc.) are required to inform ACNGT of their intention to conduct experiments that are defined as "any experiment involving novel genetic techniques that is properly contained and that can be terminated at any point without loss of containment" .

In 1988 a Genetically Modified Organism Interim Assessment Group (JAG) was established under the Environment Act. The principal role of IAG is to assess all proposals to field test or release GMOs and in so doing provide effective opportunity for public involvement in the assessment process. Private sector researchers are not bound by IAG but are encouraged to use it. In 1989 the New Zealand Government announced that a new agency, the Hazard Control Commission, would be established. This Commission, in addition to responsibilities with respect to hazardous substances, is responsible for assessing and licensing all genetic manipulation work in New Zealand and approving applications to import or release new organisms.

Among the developing countries, comprehensive national biosafety systems/guidelines have been set up in China, India, the Philippines and the Republic of Korea. Most of these guidelines are based on multitier approach, modelled on the lines of OECD and NIH and NAS (United States) biosafety procedures and guidelines for handling and release of GMOs.

In India, the Recombinant DNA Advisory Committee of the Department of Biotechnology (DBT, 1990) prepared a set of Recombinant DNA Safety Guidelines that cover all areas of research and large-scale operations involving genetically engineered organisms, excluding humans. The institutional mechanism for implementing the guidelines consists of:

- the Recombinant DNA Advisory Committee, under DBT, to formulate and update biosafety guidelines;
- institutional biosafety committees located at all centres engaged in genetic engineering research and production activities;
- the Review Committee on Genetic Manipulation, under DBT, to guide the institutional biosafety committees;
- the Genetic Engineering Approval Committee, functioning under the Department of

Environment, to review and approve activities involving the large-scale use of genetically engineered organisms and their products in research and development, industrial production, environmental release, and field applications.

In the Philippines, the National Committee on Biosafety and Biosafety Guidelines is a working group composed of representatives of the Department of Agriculture, the Department of Science and Technology, the University of the Philippines, and the International Rice Research Institute .

It has prepared biosafety guidelines that include a provision for case-by-case review of proposed releases of genetically modified organisms. A national Biosafety Committee was established in 1990 to oversee the compliance of policies and guidelines in public and private institutions. A network of 39 Institutional Biosafety Committees was created to ensure the implementation of the guidelines at the institutional level. The National Biosafety Committee coordinates with other national agencies involved in regulations, such as the quarantine services and the environmental management bureau. To date, the National Biosafety Committee has approved 15 applications for the importation of, among other things, transgenic cotton tissue and transgenic strains of rice.

BIOTECHNOLOGY RESEARCH AND DEVELOPMENT IN THE PRIVATE SECTOR

According to Sasson (1993), the development of plant biotechnologies in Latin America and the Caribbean was justified by the improvement in the capacity of manufacturing products to meet existing needs, maintenance of a competitive capacity on international markets, access to new markets, alleviation of dependence on imported inputs and its effects on the balance of payments, an orientation towards sustainable development and the conservation and use of genetic resources. Countries were advised to incorporate available biotechnologies into their productive sectors, while taking into account investment requirements, international developments and national capacities.

Jaffe (1988) analysed 21 companies in the region that were producing and selling biotechnological products. The study revealed that private companies in Argentina, Brazil,

Colombia, Costa Rica, Chile, Mexico, Uruguay and Venezuela are engaged in a commercial application of modern biotechnology for different purposes, including pharmaceutical, agricultural and industrial products. The positive results of this emerging sector are stimulating the participation of large international companies to invest in the region. Some of the leading companies in the region are Agroceres, Bioplanta and Biomatrix, Brazil; Levaplan, Colombia; Palma Tica, Costa Rica; Biotecnica, Chile; Bioquimex and Genesis Integral, Mexico; Semillas Santa Rosa, Uruguay; and Bioagro, Venezuela.

BIOSAFETY ISSUES

At present, almost 73 percent of total biotechnology research has been directed towards the micropropagation of crop species. Projects using genetic engineering account for about 9 percent of research endeavours. After more than 20 years of progressive activity in plant biotechnological research, Latin

America and the Caribbean are in a transition stage, where unsophisticated plant tissue culture techniques are expanding in many countries from the experimental to the productive stage. The multiplication of ornamental, horticultural, fruit and industrial crop species, as well as the recently growing adoption and, in some cases, development of some basic technologies such as molecular genetics tools applied to the diagnosis of diseases, genomic characterization and genetic transformation, were being developed in a smaller group of countries, such as Argentina, Brazil, Cuba and Mexico.

The release of genetically modified organisms developed in the industrialized countries has been considered to be the most important biosafety problem in Latin America. In early 1992, authorization for field testing of transformed plants was requested in the following countries: Mexico (eight requests), Costa Rica (one), Argentina (three), Chile (one) and Guatemala (one). Only Brazil, Cuba and Mexico had adopted biosafety regulations in the laboratory, all of which were based on those followed by the US National Institutes of Health. Only Argentina and Mexico had mechanisms for evaluating requests for field testing of genetically engineered organisms. Mexico had the widest experience in testing imported and locally developed transgenic plants, while relying on assistance from the U.S. Department of Agriculture. The relevant regulations were being established within existing norms on plant protection and seed marketing. At the end of 1991, Argentina set up an interministerial commission to evaluate requests for testing and releasing genetically engineered organisms. The same evaluation work was effected by an ad hoc commission in Costa Rica and Chile.

CODE OF CONDUCT FOR PLANT BIOTECHNOLOGY

Current plant biotechnology research is mainly being conducted in the industrialized world, which will have important consequences for agriculture in the developing countries. This study has shown that Latin America and the Caribbean have a relatively critical mass of institutions and laboratories to promote and develop biotechnologies to enhance agricultural production at low environmental cost.

Considering the wide availability of genetic resources in the region, the implementation of biotechnologies could, without affecting the environment, minimize the risks of dependence that have in the past exacerbated the differences between the developed countries and the countries of Latin America and the Caribbean.

One of the most controversial aspects of biotechnology is the potential danger attached to the introduction and manipulation of genetically modified organisms. Governments should reach legal and policy agreements establishing norms (codes and guidelines) to safeguard the interests and rights of individuals and to control the manipulation of genetically modified organisms during the experimental stages and subsequent industrial scaling up. Advances in biotechnologies have led to intense discussion on the need to establish rights on multicellular forms of life and their inherent genetic resources, and on how this should be done. Genetic resources and their intrinsic biodiversity are the very essence of living matter and are the raw materials for crop improvement and production. This is the main reason for the controversy over the patenting of genetic resources. The patenting of genetic resources. *i.e.* of all living plant and animal matter, which could restrict access to the genetic resources that are humanity's universal heritage.

The Fourth Meeting of the Commission on Plant Genetic Resources, held in April 1991 in Rome, Italy, agreed that the objectives of a Code of Conduct for Plant Biotechnology could be the promotion of biotechnologies for the conservation and use of plant genetic resources; unrestricted access to plant genetic resources; the promotion of biosafety to minimize risks to humans, animals and the environment; and the equitable distribution of the benefits of biotechnology between its practitioners and the germplasm donors. The role of modern biotechnologies in relation to biodiversity was discussed and argued at length at the recent United Nations Conference on Environment and Development (UNCED) in Rio de Janeiro, Brazil.

It was agreed during the meeting establishing the REDBIO, to promote the preparation and application of a Code of Conduct for Plant Biotechnology with regard to biosafety, regulations and patents.

The Workshop on the Code of Conduct for Plant Biotechnology, held in Santiago de Chile in December 1991, with the participation of 32 experts, aimed to:

1. Assess the need for, and interest in, a Code of Conduct on Plant Biotechnology with provisions for big-safety; intellectual property rights and farmer's rights; the minimization of risks; and appropriate biotechnology.
2. Discuss and coordinate cooperation with other international organizations to formulate and present a Code of Conduct for Plant Biotechnology to the governments of the region.
3. Define REDBIO's priority areas to implement the recommendations of FAO's Commission on Plant Genetic Resources for the activities defined in (i) and (ii).

The workshop participants listed the components of a Code of Conduct based on the CRGV meeting and generally acknowledged the need to establish a plant biotechnology information, advice and monitoring system. They suggested the use of FAO's Global Information and Early Warning System and of all the other mechanisms available within the framework of its global system of plant genetic resources, in coordination with other United Nations agencies. This need relates to biosafety, intellectual property, farmers' rights, socio-economic impact, appropriate biotechnology and the preservation and conservation of plant genetic resources. The system will provide free access to information to those involved in biotechnology and will develop the mechanisms for the compilation, storage and dissemination of this information.

UNCED: Biodiversity and Biotechnology

The Convention on Biodiversity and Agenda 21 were negotiated at the United Nations Conference on Environment and Development held in Rio de Janeiro, Brazil. The Convention was signed by more than 150 countries. The outcome of this conference and its implications for the region should be further examined and discussed because of the importance of these initiatives and their repercussions on the conservation and use of genetic resources.

Chapter 16 of Agenda 21 refers to biotechnology and highlights five programmes-three relevant to this paper

- the increase and availability of food supplies and renewable raw materials;
- enhanced biosafety and the development of international cooperative mechanisms;
- the establishment of mechanisms to develop biotechnology in an environment-friendly manner.

With regard to the first point, the growing demand for food will require the effective use of biotechnologies in sustainable agriculture. Modern technologies should be used together with traditional enhancement methods for example to reduce dependency on agricultural chemicals; increase the use of integrated pest and disease control systems; use organisms for the natural fixation of nitrogen; and extend the application of biotechnology to forestry. With regard to biosafety and the development of international cooperation, biotechnology will only produce its maximum benefits when there are sure methods of monitoring and control. The aim is to develop biotechnology safely, with support from international cooperation, the private sector, the non-governmental organizations, academic and scientific institutions. Efforts would have to be made to obtain procedures that are compatible with international agreements so that biotechnologies are used safely.

One of the most controversial aspects was the need for international guidelines on biosafety and on aspects related to intellectual property. The agreed text includes recommendations for international cooperation on intellectual property and innovation rights, including the rights of the plant breeder and farmer. There is also reference to the need for a guide to international agreements on safety precautions for the release of biotechnologically created products; to evaluation of biotechnology risks; and to the need for national legislation on liability and compensation.

Emphasis was laid on the need for greater efforts to increase biotechnology capacity in the developing countries, and for this to be developed and applied along environment-friendly lines. Human resource training, facilities, knowhow and economic resources were considered inadequate. It was therefore recommended that developing country capabilities be reinforced with appropriate funding. Biotechnologies should be applied particularly in developing countries with reinforced activities, research support, public awareness of the benefits and possible risks of biotechnologies, a favourable environment for investment, and promotion of scientific exchange without encouraging a "brain drain". All these initiatives should incorporate traditional methods and ethnic knowhow.

The consensus reached in Agenda 21 on genetic resources, biodiversity and biotechnology is generally in line with FAO policies. The FAO global system on plant genetic resources which includes a large section on biodiversity and many aspects of biotechnology was supported by much of the emphasis in Agenda 21. FAO's position on farmers' rights has also been reinforced by Agenda 21 with the specific call for their recognition.

7

Fish Production Sector in Transitions Period

PROPORTION OF PRIVATE OWNERSHIP

In order to enabling Hungary to join the European integration, a rigorous and occasionally painful economic restructuring has been implemented. Before the restructuring, even in 1990, more than 80% of the arable lands was cultivated by large-scale co-operatives and state farms, and only 14% by small producers. The proportion of private ownership has already reached 80%, and when the privatisation will be completed, only 2% of the arable land will stay permanently in state hands. A special element of the privatisation process is the compensation. According to the Compensation Law, the land (the area is maximised) is being returned to be original owners and/or their successors. Until the compensation process is completed, no state owned land can be sold, but the land can be rented .

In 1992 the former state owned co-operatives have been dissolved and "new type" of co-operatives were established, where the leaders were freely elected, the properties and assets went into the ownership of the members. The law let the members to decide whether they want to stay in or leave the co-operative taking their share or assets with them. About 6-10% of the members decides to leave the co-operatives and to start individual enterprise.

Those who left the co-operatives received about 5 ha land and some equipment with a value of about 4-5 thousand USD as an average. . There are about 25 "new type" of co-operatives in Hungary that are involved in fish production. The activity of these co-operatives has not changed too much, however the ownership has been restructured and they have full autonomy. Out of the 121 state farms only 25 remained in state ownership due to their strategic role in seed and breeding stock supply. Most of the state farms have been restructured and operate as share holding companies or limited liability companies, however ownership rights have been exercised by the State

Property Agency for the time of being. There have been 22 states owned fish farms in the past, and only two of them remained in state ownership. These two fish farms have special role in the Hungarian fish culture. Balaton Fish Farm is responsible for the management of the 60,000 hectares lake Balaton that is a major tourist area in Hungary, the Warm Water Fish Hatchery at Szahalombatta, that was established by the help of FAO, is a basic hatchery for quality seed supply, and also acts a training and development centre.

PRIVATISATION PROCESS

During the privatisation process, various size and type of fish farms have been and being established. Most of the private fish farms are fairly small in area, but there are private fish farms that operate several hundred ha fish ponds. There are considerable changes in the legal, financial and institutional background as well, in order to facilitate the transition process into market economy. Numerous laws and acts have been passed in the past period such as: Compensation Law, Law on the Mandatory Restructuring of the State Owned enterprises, Law on Co-operative Transformation, Market Regime Law, to mention only the most important ones. Although the new Fisheries Act has not been fully elaborated yet, a recently issued Ministerial Order brought some considerable changes in the application of the fisheries rights. According to this regulation, all companies, organisations and private entrepreneurs have equal right to apply for the fisheries right of a water area owned by the state. The Ministry gives priority to those applicants that have high professional skill and present a well elaborated plan for the fisheries utilisation of the given water body. In order to promote private enterprises various governmental interventions have been implemented in form of reduced taxes special loans and interest rates. A special Agricultural Fund has also been established recently to promote the restructuring process, to improve the technical conditions of agriculture production, and to preserve the gene stocks and natural resources.

One component of this special fund is the Fisheries Fund that provides financial support for the rehabilitation and fish enhancement of natural waters, and it also support research on this area. However other components of the Agricultural Fund can also be used by the fish production sector for the maintenance of high value fish breeding stocks and for the technical improvement of production facilities. Financing is one of the main problems for fish farmers. Although the bank systems has been changing, special agricultural loans are not readily available, running times and interest rates are not tailored to the special conditions of the agricultural production.

Difficulties in financing would create an even more serious liquidity crisis in fish production, where the production cycle is long, and it takes three years to grow the fish up to market size in traditional fish pond production. The respond of the farms to the increasing financial difficulties was the cutback of the production. In the centrally planned economy trade was regulated administratively rather than by the market. The producers were dependent on the various state owned trading companies that were in a monopolistic position having exclusive export-import licence.

MARKET REGIME

Although foreign trade has been "liberalised" in the early eighties, business contracts and knowledge on this field remained the privilege of a fairly small number of experts. After being directed or guided by the state, most of the producers have not been prepared for the new challenges, they lack marketing skills, business and market information. Thanks in the entrepreneurial spirit of the Hungarian farmers, most of them are adapting well to the new situation.

The new law on "Market Regime" helps to regulate the market and to secure free and fair competition for all participants on the market. According to this new law, so called "Product Councils" have been organised with the participation of the producers and traders. One branch of the product councils is the Fish council, that suggests limit prices, contingents, export-ban periods etc. Under the previous regime, technology transfer was accomplished through the technical staff of agricultural scientist, economists, and veterinarians employed by the state farms and co-operatives, however this functional system has been disappeared during the restructuring of the agriculture. The reestablishment and strengthening of an effective extension system has been started, however it is not functional yet. There are some firms and groups which provide private agricultural advisory services to farmers, however the main constraints confronting the establishment of a flourishing extension services by these new practitioners are given as: lack of money to pay for the service, in some areas community prejudice against former employees of the cooperatives and state farms, and farmer confidence in his ability to solve his own problems .

Larger scale farming units such as many privatised state farms and larger co-operatives continue to employ their own technical advisory staff or purchase the required assistance at full cost from private extension services, universities or research institutions, but small scale farmers are still facing with difficulties to get access to special services and information. No extension service is available for fisheries and aquaculture. However the Fish Culture Research Institute and other institutions that are involved in research and development on this field, are providing assitance and services through various channels. The Association of Fish Producers also assists the technology transfer providing linkage between farmers and research institutions, organising meetings and training programmes. Agricultural education and training is undertaken by a number of institutions including three agricultural universities in Hungary.

Although no MSc courses on fisheries are available at these universities, opportunities exist for students to be specialised in fisheries and aquaculture, an there are also state scholarships to study these subjects abroad. Vocational training and education is limited due to the relatively small demand by the sub-sector, and courses and post graduate education are held intermittently. There is one special school in Hungary for the training of skilled fishermen. The regular 3-year course caters for 20-25 students annually, but post graduate master-fishermen courses are also held occasionally according to the demand. There is a solid research and development background in aquaculture and fisheries in Hungary. Special methods in the propagation and pond production of common carp that were elaborated by Hungarian scientists are widely applied world-wide. The core institution for research and development is the Fish Culture Research Institute (HAKI) at Szarvas, that is an internationally acknowledged centre in this field. In spite of the serious cutback in financing research during the past three years, the Institute could preserve and even develop the valuable research capacities as a result of an internal restructuring programme.

PROBABLY CONSISTS OF FISH FROM A MIXTURE OF DISTINCT POPULATIONS

When an oceanic salmonid catch is brought ashore, it probably consists of fish from a mixture of distinct populations from different river sources. We have already seen how salmonids can be genetically very divergent between rivers and these differences are likely to be reflected in factors such as growth rates, age at maturity or other life history parameters. If the fishery is managed as a single unit, there is the potential for the overexploitation of some populations, or the suboptimal harvesting of those that are more abundant. In order to enable proper management and obtain the maximum sustainable yield from such fisheries, we need to estimate the proportions of the various populations in the fished resource. If the contribution of the various populations varies geographically or over time, then fishing effort can be directed to exploit the strongest populations, while pressure on weaker populations can be reduced, by targeting fishing to particular regions or particular times.

Early attempts to determine the relative contributions of individual populations to a fishing stock by tagging or branding fish as they departed from specific rivers were relatively ineffective because of the high initial labour costs and the loss of markers in a proportion of the fish.

Far more effective is the use of genetic markers. Allozyme markers have been widely employed in the management of Pacific salmon, using comparisons between allele frequency data for the individual river populations and allele frequencies in the total catch. Of course, to make it work, the differences between river populations must be sufficiently great that allele frequencies at minimally a few of the loci scored will enable identification of that population in the total catch. Fortunately, there are statistical methods available which made these comparisons possible even when allozyme allele frequency differences are not very great. However, the potential resolving power of microsatellite markers is much greater than allozymes, and they are now the markers of choice for all studies of MSA.

CONSIST OF A MIXTURE OF NATURAL AND INTRODUCED STOCK

MSA is also relevant to some sport fisheries that consist of a mixture of natural and introduced stock. The introduced element is either transplanted from another area or derived from hatchery production. A good example is the trout, *Salmo trutta*, which exists in Europe in two different forms. One, the resident or brown trout, inhabits rivers and lakes and never goes to sea, while the other, the sea trout, is anadromous and, as its name suggests, migrates into the sea and spends some time there before returning to its natal river to spawn. Where waterfalls that are impassable to trout are present in rivers, the populations above the falls will be resident trout and those below will be sea trout. Because sea trout grow bigger than resident trout, they make a better fighting fish for rod and line and, mainly for this reason, restocking of river populations above impassable falls is often carried out with hatchery-reared sea trout. The expectation was that there would not be a problem with the survivors of these introduced sea trout breeding with the natural population of resident trout because, once they had migrated to the sea, they would not be able to return to their original position (above impassable falls) in the river to breed. However, allozyme and microsatellite-based studies in various parts of Europe have shown that this does not always happen. It is clear that some stocked sea trout do not migrate to sea but remain in the populations of resident trout with which they will then breed. This leads to a loss of the unique genetic identity of the resident stocks and is an important conservation issue.

In North America, the cutthroat trout exists as a complex of up to 15 different recognised subspecies, and there has been extensive introduction of the different subspecies into non-native habitats. Genetic marker analysis has demonstrated that this has resulted in an extremely complex pattern of hybridisation and introgression. This can be bad news because the hybrids often suffer developmental abnormalities that make them less fit than the pure-bred subspecies. Furthermore, as in the case of brown trout in Europe, there is evidence of genetic swamping which threatens the uniqueness of locally adapted populations.

Stocking of rivers for salmon fishing using non-indigenous fish has been extensively carried out across the world. Some rivers are thought to contain no descendents of the original indigenous populations.

Such eradication of the native population was believed to be the case in many Danish rivers until analysis of microsatellite data from archived scales was used to identify indigenous salmon. DNA was extracted from adipose fin samples from recently caught fish from various Danish rivers and from old scale samples (archived between 1913 and 1954) from some of these rivers. Samples were scored at six microsatellite loci and analysed using a 'self-classification test' that enabled individual fish to be 'assigned' to particular populations based on their multi-locus microsatellite genotypes. Information about the microsatellite alleles and their frequencies in indigenous river populations were obtained from the scale samples which pre-dated the extensive introductions of non-indigenous salmon. In spite of these extensive introductions, results revealed that around 11% of fish currently in the rivers had genotypes that were characteristic of the indigenous fish. This showed that descendents of indigenous fish were still present and identifiable and had not become genetically altered or homogenised by extensive hybridisation with exogenous fish.

Using non-invasive DNA sampling of salmon (small fin clips or blood samples that do not harm the fish) to identify indigenous individuals, it has been possible to undertake breeding programmes for re-stocking rivers with native stocks. This study demonstrates the power of microsatellite variation as a tool for MSA that can identify indigenous fish and allow restoration of locally adapted stocks. Now we will consider how population size can impact on the genetics of organisms. The numbers of individuals within a population that contribute to the next generation is a key factor in the maintenance of genetic variation, and there are two situations where this will be of relevance. Firstly, small population sizes can occur in the wild due to natural physical (*e.g.* climate extremes) or biological (disease) forces and also due to anthropogenic effects such as pollution or overexploitation. Secondly, small population sizes are a necessary element of aquaculture that involves the use of a hatchery to manage the production of young fish and shellfish. Therefore, the genetic concepts and constraints focused on the conservation of scarce species, or small endangered populations within a species, are very similar to those that are relevant to hatchery production in aquaculture.

The concept of random genetic drift, that is the process that causes random changes in allele frequency from generation to generation. When the effective population size (N_e) is very large, changes in allele frequencies between successive generations will be very small. On the other hand, in populations that have a small effective size, there will be a large variance between generations in the frequencies of alleles, that is natural fluctuations in allele frequency between generations will be much greater in small populations. In such populations with small effective size, this will lead to a reduction in genetic diversity over time that can be identified as a loss of alleles and loss of heterozygosity. Why does it matter that alleles might be lost? Each variant allele at each coding locus in a population can be regarded as part of the 'genetic resource' of that population. An allele alone, or in combination with other alleles or loci, could be responsible for conferring on its carrier a valuable trait such as increased resistance to a particular disease, better cold tolerance or faster growth.

Therefore, the loss of any allelic variants is a potential loss of valuable genetic resource. Of course, if most allelic variation at coding loci is neutral, then this is less important, but we would be unwise to ignore the certainty that at least *some* variants at coding loci will be advantageous. If not now, then most likely in the near- to medium-term future, global warming will bring about the increasing importance of high-temperature-resistant allelic variants at biochemically important loci in temperate aquaculture species. Such alleles may be effectively neutral until extreme summer temperatures reveal their value. A second important consequence of small effective population size is the phenomenon of inbreeding that is brought about through matings between closely related individuals. From the genetic perspective, inbreeding increases homozygosity and almost always has deleterious phenotypic effects, making inbred offspring less likely to survive than non-inbred offspring. The phenotypic consequences of inbreeding (low viability, poor growth, abnormalities) are labelled inbreeding depression.

METHOD OF SINGLE-STRAND CONFORMATIONAL POLYMORPHISM

The method of single-strand conformational polymorphism was used to identify mtDNA haplotypes from the same 15 populations of European flat oyster as had been used for allozyme and microsatellite analysis. Single-strand conformational polymorphism involves denaturing polymerase chain reaction (PCR)-amplified DNA (in this case a 313-bp fragment of the 12S-rRNA gene) at 95æ%C before running the product down an acrylamide gel. The final position on the gel of the single strand DNA is identified by ethidium bromide. A total of 14 haplotypes were identified. One haplotype predominated in the Mediterranean, while another prevailed in the Atlantic populations. A third haplotype was common in samples taken at the geographic extremes of the species distribution. So the mtDNA analysis generally supports the idea of an Atlantic stock and a Mediterranean stock of flat oysters.

However, unexpectedly, some genetic similarity is shown between Norwegian and Black Sea oysters and this is difficult to explain. The overall pattern of isolation by distance is much more strongly demonstrated by the mtDNA data than with allozymes and microsatellites. The average F_{ST} is 0.244, an order of magnitude greater than that for microsatellites.

How can this difference be explained? For most animals, because mtDNA is haploid and maternally inherited, there is a (theoretically fourfold) smaller effective population size compared with diploid nuclear markers. This means that mtDNA variation is likely to be four times more sensitive than nuclear markers to factors that reduce population size and create bottlenecks (such as overfishing, habitat loss or disease infestations). However, in flat oysters, this would be reduced to twofold because they are sequential hermaphrodites and individuals can be both male and female in their reproductive lifetime. There are two other factors that might possibly contribute: unbalanced sex ratio and/ or variance in female reproductive success. Flat oysters usually develop first as males and only in later years change to produce female gametes.

Within any particular season there is generally a 3:1 male to female ratio. This deviation from the 1:1 male to female ratio is probably exacerbated in areas where there is a high prevalence of the parasite *Bonamia ostreae,* because highest mortalities from this parasite usually occur in 2–3-year-old adults. These older oysters are more likely to be female than male, which increases the proportion of males in the genetically effective population. An unbalanced sex ratio will reduce effective population size. Variance in individual reproductive success is large in organisms, like oysters, that produce huge numbers of eggs and such variance is thought to be greater in females than in males.

This variance between individuals could be an important component in reducing effective population sizes of oysters. So there are some reasonable biological explanations for the magnitude of difference in estimations of population differentiation based on mtDNA and microsatellite data. Populations at the extremes of the geographical range of a species are likely to experience stronger environmental pressures and greater temporal variability in reproductive success than populations well within the species range. This is expected to result in an increased likelihood of bottleneck events and overall smaller effective population sizes.

We can see the genetic consequences of this in these studies in oysters: at allozyme loci, microsatellite loci and in mtDNA markers, the most northern population from Norway and the most eastern population from the Black Sea both have generally less genetic variation (fewer alleles and lower heterozygosity, fewer haplotypes) than other populations. Overall then, we can see how different genetic markers can throw up different estimations of population differentiation and that the analysis and interpretation of data from these different markers must be carried out with reference to the biology of the organism.

DEVELOPING A GREEN-FIELD SITE

Often developing a fishery from scratch will require the development of a green-field site. However, before any development takes place a suitable site must be found. In many respects this can be the most difficult aspect of the entire procedure of developing and creating a fishery.

A fishery needs land and adequate good-quality water. The UK is a very densely populated island. In the past, its freshwater resources have been adequate for supporting the population with drinking water, and water for agriculture, industry and recreation. However, even in the past, stresses on the resources were such as to result in severe pollution and local shortages in supply. Now, with changing climate and growing population pressures, the quality and quantity of water is constantly under pressure in parts of the UK as rainfall appears to be declining and becoming less predictable. Whatever the cause and the degree of impact that this will have on the environment, it looks certain that the current trend towards heightened water awareness by the general public is set to continue for the foreseeable future. It is this awareness that will drive those organisations responsible for water regulation to be increasingly vigilant when considering applications for abstraction and discharge consent. The need for continuity and protection for the potable supply carries immense political weight.

Obtaining a plot of land suitable for development is not all doom and gloom, far from it. As this is still a wet island, standing waters occur frequently in lowland areas. It is, however, probably fair to say that any future developments will need to have an increasing regard for their potential impact upon the environment.

The bureaucracy involved in turning a green-field site into a fishery is considerable; however, before considering this it may be pertinent to consider some aspects of locating a suitable parcel of land for development.

FINDING AND IDENTIFYING A GREEN-FIELD SITE SUITABLE FOR A FISHERY

Before identifying a suitable site and obtaining some form of tenure (licence, lease or purchase) on it, one should consider exactly what is needed, both in terms of what is essential and what is desirable in the site. Ultimately the type of fishery that is to be developed will be the main criterion for deciding the suitability of the site, although there may be some flexibility to modify the type of fishery to suit the site. For example, certain water-quality criteria are vital in determining whether a fishery would be suitable for trout. Certainly water quality is not a factor that should be compromised: generally if it is not suitable to begin with, in practice it will be difficult to improve at a later stage.

Finding the right site in a convenient location is very important. Only investigate sites within a predefined radius of your base. Consideration should be given about what is a reasonable distance to travel on a daily basis. Of course, this decision is a matter of personal choice, but it stands to reason that, as most people live in the urban environment, then most will have to travel to rural areas to find and locate a green-field site. There is little point in coming to an agreement to develop a site many miles from home only to find that the strain of travelling is too great to sustain. The actual finding of a potential site requires a great deal of personal effort. It is a very time-consuming procedure.

Although some fishery owners will surely report that they just ' came across ' their site, most sites are found after a great deal of time and effort. Time and effort are required because there is nowhere that one can go to find out where a defined green-field site awaiting development is situated. There are many ways of obtaining information that will make the task of finding such a site easier. These may include speaking to local people; success may often depend on word of mouth and who you talk too. However, local knowledge should not be overlooked. There can be no doubt that the best method in condensing the search area is the use of Ordnance Survey maps and the ability to read and understand contour lines. If this is coupled with a map showing geological and hydrological features, much of the initial detective work can be done from the comfort of your own home. The Ordnance Survey's 'pathfinder' 1:25,000 series of maps is well suited. They are available from bookshops within the local area to be investigated. The area should be near to a water supply from either running or spring/groundwater sources; it should have an impermeable substrate and be accessible. If the area broadly fits these criteria then it will merit further investigation, and it is time for a site visit. On-line searches using geographical engines such as Google Earth or Google Map can also help refine the search and clarify the setting around an identified area. Many of the classic signs of a suitable site will only ever be found by close observation in areas of potential site development. These may include the following:

- standing surface water;
- boggy appearance;
- *Juncus* (soft rush) and *Carex* (pond sedge)-type grass tufts or other aquatic plants;
- nearby water source.

SITE ASSESSMENT OF A SUITABLE GREEN-FIELD SITE FOR A FISHERY

Once a potential site has been located, a decision needs to be taken about the site's suitability for the proposed use. To decide this some further information needs to be drawn together for consideration. It is suggested that the following factors be considered when assessing a green-field site for its suitability for a particular type of fishery:

Site Security

The potential security of the site is easily overlooked. There can be little point in successfully creating a fishery only to see the fruits of one ' s labours disappearing by theft (poaching) and predation (herons, cormorants, mink, etc.). An isolated location will make the security of the site very difficult. The theft of fish is widespread, even from angling waters where fisher-men are in attendance. A near-ideal site might of course have accommodation on site or very nearby. Every site is unique. It is important to consider site and stock security as one of the contributing and deciding factors in a decision about suitability for the purpose.

Site Availability

There is little point in formulating plans to develop and construct a fishery if the potential land is not commercially available. This needs to be confirmed at an early stage. Local knowledge can be an advantage: a question at a local pub, or to a passer-by, as to the ownership of the potential plot of land may be helpful. It is advisable to keep the amount of information departed to a minimum: enquire the name of the owner without giving your reason for wanting the information, other than to contact them. It is not unusual to find a piece of land that the local inhabitants cannot identify an owner for, as often with suitable sites they are poor agricultural land that is left fallow for year on year. The local farmer may well be your next best option. Do your research first as they might turn out to be the owner. An additional source of ownership information is through the Land Registry.

Access

The site will not be viable if access makes either the construction or the future management of the facility difficult and expensive to undertake. It is advisable to take a very detailed look at this aspect early in the site visit. The laying and maintenance of a hard surface suitable for vehicular access can often be a very limiting factor both logistically and financially.

Nature Conservation

Increasingly, many parts of the countryside are being protected by various conservation -status designations. It is important to establish as early as possible if any of these designations affect the potential site, or indeed if any are up-or downstream. The designation can affect whether development would be allowed and, if it were, whether the management regime would be required to coexist with the conservation status.

The development of an area that has been designated with a special status will as a minimum undoubtedly involve outside bodies such as national and local conservation bodies. If the area is designated as a Site of Special Scientific Interest (SSSI), then Natural England will need to be consulted.

Soil types

Fortunately, the tell-tale signs that may lead to a site investigation in the first instance can also be indicators of the suitability of the soil for holding water, namely visibly wet and swampy ground. Fisheries rely on a natural bottom and sides to the pond or lake, therefore the material that is left after construction must be impermeable to keep seepage to a minimum. The suitability of soil for a fishery is more involved than just its ability to retain water. The soil richness will directly reflect in the fertility of the pond and thus its ability to support a balanced aquatic ecology, including a fish community. The best way to establish suitability is to perform a soil survey.

Soil surveys should be done as early as possible. Soil sampling can be carried out in two ways, which are dependent on the analysis required from the sample. They are:

- disturbed samples;
- undisturbed samples.

Disturbed samples do not represent exactly how the soil lies in its natural state (before sampling), whereas an undisturbed sample represents exactly how the soil lies in its natural state.

The disturbed sample is used for the simple tests: those that can be performed at the site while making the initial site visit. The disturbed-sample method also includes a simple test, which can be done at the site to establish soil texture and its ability to retain water.

It follows that the undisturbed sample is used for the more sophisticated tests, which are more suited to laboratory investigation, such as chemical analysis. The undisturbed sample requires a more involved procedure to obtain it. Before developing a green-field site for a fishery, it is suggested that both types of sample will be required and the appropriate analyses performed.

The initial survey may be termed the ' reconnaissance survey ' . It should be quick and provide a general idea of the soil varieties present and where they are found. A survey like this should be conducted by digging several open pits and then examining the exposed soil profile. From this it should be possible to determine what parts of the site are suitable for pond construction (those having good impermeable soil), and those parts that are not suitable (gravel outcrops, thick layers of organic material, etc.).

Close inspection of the open pit will reveal the soil profile. This comprises the soil horizon (layers of different types). It will be worthwhile making a sketch or taking a photograph, and taking a sample of each layer. This is not quite the same as an undisturbed sample, as the soil is only lying in its natural state in the horizontal plane.

It is worth noting here that should soil be required to construct a dam or dyke, it needs to have the cohesive properties or plasticity index of 15. Soils that contain too much sand or gravel should not be used because they will not hold water. Clays and silty clays are best. Soils that contain at least 20% clay by volume are suitable for making impounding banks.

It is clear that a soil ' s ability to retain water and to act as a nutrient base is important to the creation of ponds and lakes for recreational fisheries.

REASONS FOR PRODUCTION GROWTH

However, the CAGR for other leading fish-producing coastal states has been 4.85 per cent for West Bengal, 4.47 per cent for Tamil Nadu, 3.37 per cent for Gujarat and 3.25 per cent for Odisha between 2004-05 and 2008-09, it said. Interestingly, non-coastal state of Haryana had registered the highest growth rate of 14.7 per cent in fish production followed by Chattishgarh (11.29 per cent) and Uttar Pradesh (7.05 per cent), the industry body said. The factors that have contributed to the development of world capture fisheries over the last half century, a period that witnessed a rapid growth (circa 6% p.a.) in landings during the 1950s and 1960s, but a much slower rate of increase in the subsequent three decades. Indeed, as we saw earlier, in some sea areas there has been a leveling off or decline in catches for certain groups of species (*e.g.* cod from the NE Atlantic), mainly due to stock depletion caused by excessive pressure of fishing. In our discussion, therefore, we will consider how various factors (some positive, some negative) have influenced global fisheries production.

Government and Donor Support

While government and donor support to the sector has historically taken a variety of forms, such support can fall into one of three categories. Direct payments such as price support payments, grants towards the purchase of new fishing vessels and/or gear, compensation payments for disrupted fishing seasons, and disaster relief payments etc.

These enhance the revenues of the recipients and are paid from government or donor budgets directly to the fishers. One example of such support was the September 2006 Irish government announcement of almost €56 million of government and EU grant aid for the introduction of 38 new vessels into the Irish Whitefish fleet, including a special €3 million scheme designed to assist young fishermen to obtain their first vessel.

Cost reducing transfers such as fuel tax exemptions, subsidized (or free) ice, gear and vessel provision, preferential rates of interest for vessel/gear purchase, and reduced charges by government agencies etc.

These are paid by the government or donors to fishers to reduce capital and/or running costs. Over-zealous aid agencies, for example, have contributed to the pressure on fish stocks in the eastern Indian ocean by supplying too many new boats to coastal communities hit by the 2004 tsunami. While Thai geo-informatic agency estimates indicated that 4700 fishing boats were destroyed across 406 Thai fishing communities, Thai Fisheries department calculations suggested a total of 24 400 boats have been replaced (the majority of which had been donated) or repaired!

General services refers to all other support schemes, such as the construction of port facilities, research expenditure, restocking costs, market intervention schemes, protection of marine areas, regional development grants, managerial and technical support to fisher organizations etc. These benefit the industry, but the costs are not fully recouped from the sector. One illustration of this was the modernization of the Cotonou fishing port in Benin in 2004. Costing US$7.3 million and benefiting some 4000 fishermen, the costs of the scheme were largely met by the Japanese International Cooperation Agency (92.38%) rather than the fishermen themselves.

While estimates as to the annual level of support vary widely—the OECD mentions a level of around US$6 billion across OECD member states, and the World Bank suggests a global figure of between US$14 and 20 billion—what is undeniable is that much of this support, by reducing the private costs of fishing activities, has encouraged fishers to trade up vessels and/or gear and expand fishing activities with a consequent deleterious impact on stock levels. A belated recognition of this, allied to the precarious nature of many of the major commercial fish stocks, now sees support being (partially) reoriented towards the decommissioning of vessels, the buy-out of licences and permits, the introduction of re-training schemes and a raft of other measures designed to reduce the pressure on fish stocks.

The Attraction of Fishing

Fishing has proved a relatively attractive commercial and/or subsistence occupation for new entrants in a number of instances, notwithstanding the widely-held thesis that fishers are among 'the poorest of the poor'.

Reasons for this are manifold. In Panama, for example, the mechanization of sugar and rice production led to a massive influx of displaced agricultural wage-labourers into the artisanal shrimp-harvesting sector between the 1960s and mid-1980s.

The Chilean devaluation of the early 1970s attracted substantial inward investment into the industrial fishing sector, with one beneficiary claiming to have recouped their US$200 000 investment within the space of a week. A reversal of the anti-export bias in Argentine macro-economic policy, allied to tax exemptions on new fishing vessel purchases and a relaxation of the rules governing entry into the industrial fishery, facilitated a dramatic growth in the Argentine offshore fleet in the 1990s.

In Canada, the availability of insurance under the Special Seasonable Fisherman's Benefit Programme accounted for the presence of one-half of the inshore fishing boats in Newfoundland, and postponed a restructuring of the industry until the 1990s. All the above examples do share one thing in common, however. In each instance, fishing—given the prevailing technology—is more attractive than alternative employment and/or investment opportunities, and so entry into (or exit from) the industry is encouraged (discouraged). While the relative attractiveness of the industry can be modified by shifts in government/donor policies and access regimes (and in the longer term by technical change), linkages between the sector and other labour markets and investment opportunities (at the local, national, regional and global levels) can also have a material effect upon fishing effort and consequently, stock sustainability.

The Property Rights Regime

A third supply-side factor encouraging the over-exploitation of fisheries resources is the nature of the underlying property regime governing access to fish stocks. While land is relatively easy to demarcate, territorial rights at (and over) the sea are less easy to define/enforce and so, by default, open access fisheries were historically the norm rather than the exception. This principle was formalized by Hugo Grotius in 1608 who coined the phrase *Mare Liberum* the 'Freedom of the Sea', which has dominated humanity's approach to fishery resources for the last 400 years.

In open access fisheries, serial depletion of species is routine. The same tendency is also apparent under common pool property regimes—where resources pertain to a defined community—members seeking to maximize their gain by harvesting more (and more) fish. The 'race for fish' that develops encourages over-investment and results in the progressive over-fishing of fish stocks—with species only escaping over-exploitation if their range extends beyond the scope of current fishing technologies—an increasingly finite option, and/or the open access regime is terminated. Since the mid 1980s, increasing attention has been orientated to property rights regimes in fisheries, and the idea that the assignation of private property rights can resolve the problem of over-fishing. Current research, however, suggests that the allocation of private property rights via quotas may not necessarily resolve overfishing problems, but could in fact foster the concentration of capital and an increase in by-catch.

The Technological Revolution

An important supply-side driver has been the developments in fishing technology. While the switch from sail to steam power and the introduction of hydraulic landing gears permitted the deployment of large otter trawls and seines prior to the First World War, the pace of technological development in the sector has accelerated over the last half century. Echosounders and netsondes were developed for use on vessels and fishing gear in the late 1940s and 1950s, enabling fishers to locate—and then position fishing gear with more precision over—schools of fish. More recent acoustic innovations, such as the multi-frequency towed device developed by CSIRO have, by allowing scientists to distinguish between both species and sizes of fish, permitted the more accurate assessment of the biomass of species such as the orange roughy and blue grenadier, and it is probably only a matter of time before such devices are deployed across the industry so as to further enhance catches.

The introduction of synthetic fibres in the 1950s not only permitted the development and rapid dissemination of power block seining, but made mid-water trawling and gill-netting (a common fishing technique across much of the developing world) more effective, while underpinning the emergence of modern long-lining techniques. Fleet improvements have also contributed to an increase in fishing power, with relatively high-cost steam-driven trawlers being crowded out of many fishing grounds by diesel-engined trawlers with cheap hydraulic winch systems by the 1970s. Improvements in hull shape and vessel design have proceeded apace. The deployment of navigational aids such as Global Positioning Systems (GPS) has allowed the location of fishing areas or topographic features to be pinpointed with greater accuracy and the whereabouts of vessels to be determined within tens of metres. GPS, in conjunction with digital charts and sidescan sonar, enables more detailed seabed mapping and the construction of highly detailed three-dimensional models of fishing grounds. Incorporating information on time and tides, weather conditions and past catches further improves the effectiveness of fishing operations. Post-harvest developments, in particular improved freezing and at-sea processing technology (case of industrial fisheries) and the more widespread distribution of ice-blocks (artisanal fisheries) have also enabled fishers to extend the range and duration of fishing trips.

Integral to these developments has been scientific research. Growing knowledge about fish behaviour has been complemented by ever more precise measurement and calibration of gear performance. In the 1960s commercial gear development research led to the development of more efficient trawling techniques and fish/shellfish selection. The impact of fishing gears on the sea-bed, twine thickness and improving the energy efficiency of fleets in the wake of the 1973 oil crisis came to the fore in the 1970s, with the development of species and size-selective commercial fishing techniques being given a high research priority in the following decades. These technological developments have facilitated both the intensification and extensification of fishing, increasing the efficiency of fishing and expanding the areas that may be fished. Presently, for example, modern long-lining gear generally comprises a 60-80 mile main-line and up to 3000 baited hooks suspended off a series of branch-lines, 40 mile drift nets are commonplace, while the maw of the Gloria supertrawl net is large enough to accommodate twelve Boeing 747s.

While technological change has been a dominant force in the expansion of world capture fisheries, it has also almost certainly contributed to the overexploitation of many fish stocks and in some cases led to their near collapse. The likelihood of this happening has been greater in fisheries where the regulatory controls have been weak or ineffective, and where increases in the fishing power of vessels have been allowed to take place without regard to the consequences this might have for the effective catching capacity of the fleet. The post-war development of the NE Atlantic herring fisheries provides a salutary illustration of this, with the rapid adoption of new fishing methods (notably pelagic trawling and purse seining) during the 1960s resulting in rapidly rising catches but, eventually, to stock depletion and the closure of the fisheries in the North Sea and West of Scotland. That said, not all technological developments have been to the universal detriment of fish stocks. The introduction of vessel monitoring systems (VMS) has aided regulatory authorities to ensure skippers comply with national/international fisheries regulations, although their use has not yet gained worldwide acceptance. Equally, the growing sophistication of fishing gears has helped reduce by-catch and the ensnarement of non-target species.

Changing Consumer Tastes

An increased attention to diet and personal health over the last half century or so has also stimulated consumption demand for fish. Fish is low in saturated fats, is a rich source of high-quality protein and Omega-3 fatty acids, and also provides essential minerals (iron, calcium, zinc—from shellfish—iodine and selenium) and vitamins (B and D) which are easily absorbed by the body. Omega-3 fatty acids are now known to be particularly beneficial in terms of cardiovascular health, for example, and some studies have indicated that early exposure to Omega-3 fats may also enhance brain and retina development. The Iowa Department of Public Health in USA, for example, suggests that eating fish may protect against a variety of diseases and illnesses in adults, including cancer, dementia, diabetes, depression, rheumatoid arthritis, psoriasis, prostrate cancer and autoimmune disease.

While information and advice like this has undoubtedly stimulated the demand for fish products—a demand which was aided by the relative historic cheapness of the product compared to alternative protein/food sources such as red meats—it has been tempered in recent years by public health concerns regarding the levels of environmental contaminants encountered in fish/shellfish. Some farmed salmon and fatty fish have been found to have significantly high concentrations of PCBs (polychlorinated biphenyls) and dioxins, while high levels of methyl mercury have been encountered in predatory marine (tuna, swordfish, shark and halibut) and large freshwater fish. As a consequence, many food standards agencies in the developed world have issued advice on what they consider to be 'safe' fish intakes. In the developing world, consumption of raw or inadequately cooked fish can also pose serious threats to health, with medical complaints such as bile duct cancer, gallstones, diarrhoea, and peptic ulcers increasingly being attributed to fish-borne zoonotic parasites in Southeast Asia.

Although tastes have not changed in China and Southern Asia, where fish has always been a consumption item of choice, rising incomes in the region have nevertheless also contributed to growing consumer demand. Currently, global fish consumption stands at around 15 kg per capita—double the level of thirty years ago—and has been instrumental in underpinning the growth and internationalization of fish commodity chains. The international fish trade grew nine-fold in the twenty years from 1981-2001 (from US$6.1 billion to US$56 billion), for example, with a particularly sharp rise in the receipts accruing to developing countries (up from US$3.4 billion to US$17.4 billion) over the same period. Moreover, according to the same author, this growth in international trade has not had a detrimental effect on food security in the form of fish for food—although the same cannot perhaps be said for fish stocks.

RECENT TRENDS IN FISHERIES PRODUCTION

THE REGIONAL PERSPECTIVE

Asia

Asia is the world's foremost capture fishery and aquaculture producer. In 2004 the regional marine harvest alone totaled 32 million tonnes of marine fish, 16 million tonnes of molluscs, 14.4 million tonnes of aquatic plants and 5.9 million tonnes of crustaceans. Although the growth in marine fish landings since the early 1990s has been rather modest (up 23% from around 26 million tonnes) due to the gradual (over) exploitation of stocks, crustacean extraction has doubled and the harvesting of molluscs and aquatic plants has trebled over the same period. The principal marine fishing nation in the region is China, although a further nine states landed in excess of 500 000 tonnes in 2004.

The Table, nevertheless, fails to illustrate the profound re-alignment in regional production that has taken place since 1990. In that year Japan landed 8.1 million tonnes (30.7% of regional landings) and the Republic of Korea 1.8 million tonnes (6.8%). However, a series of fishing capacity management plans introduced in the two countries during the intervening years has seen capacity—and with it marine landings—sharply reduce in both countries. In contrast, production has soared in most of the less developed economies (Chinese landings up 147%, Vietnamese up 132%, Indonesian +92%, Myanmar +84%), India being somewhat of an exception (production up 23%) due to its principal marine fisheries already being rather closer to full exploitation.

The principal molluscs extracted are the high-value Pacific cupped oyster-*Crassostrea gigas* (4.3 million tonnes in 2004) and Yesso scallop-*Patinopecten yessoensis* (1.5 million tonnes), and the lower value Japanese carpet shell-*Ruditapes philippinarum* (2.9 million tonnes). While China dominates mollusc production in the region (61% of reported 2004 regional harvest), Japan (1.8 million tonnes, 11% of regional harvest), the Philippines (1.3 million tonnes, 8%) and the Republic of Korea (800 000 tonnes, 5%) also possess significant mollusc-extraction sectors. The main aquatic plant harvested is Japanese kelp, the harvest of which doubled to just over 4.6 million tonnes during the period 1990-2004, although regional production of *Wakame* (a kelp-like plant), laver (*Nori*) and Zanzibar weed have also surpassed the million tonne mark (2.5 million, 1.4 million and 1.1 million tonnes respectively in 2004). The main driver in aquatic plant extraction is, once again, China (76% of 2004 regional harvest). Although prawns/shrimp dominate under the crustacean category

(73% of category volume in 2004), serious viral pathogens affecting the region's main indigenous *penaeids* (the Giant tiger prawn and the fleshy prawn) constrained growth for much of the 1990s. The introduction of the whiteleg shrimp at the turn of the century however promptly restored momentum to the sector, with this one species alone providing a harvest of 710 465 tonnes (17% of regional shrimp/prawn harvest) in 2004. China supplied 53.5% of the regional crustacean catch the same year.

Regional production from inland water fisheries has grown more than three-fold since 1990 to 27.4 million tonnes in 2004, although the majority of this (86.4%) arises through aquacultural operations. China dominates once more (67.8% of 2004 harvest), having increased landings from 4.9 to 18.6 million tonnes over the same period. India (2.95 million tonnes), Bangladesh (1.5 million tonnes), Indonesia and Myanmar (824 000 tonnes apiece) and Thailand (531 000 tonnes) also post significant inland landings. Carp—chiefly Common, Crucian, Grass and Silver—account for two-thirds of the regional inland fisheries output.

FAO calculations suggest that aquaculture production provided an income in 2004 of around US$27.9 billion for participants, up from US$7.3 billion in 1990—with the main beneficiaries currently being China (US$23 billion) and India (US$2.2 billion). Per caput food supply varies across the region being highest in East and Southeast Asia (26.8 kg), and markedly lower in both Southern Asia and the Near East (both 5.6 kg). Per caput supply is particularly high in Japan (64.7 kg), where the tradition of eating fish is very strong and fish is generally more important than meat in the diet, while the converse holds true in Afghanistan (zero kg). In Bangladesh, Indonesia, and Sri Lanka fish provides more than half of the daily animal protein requirements.

Europe

In Europe (including the former USSR and the new transition economies) marine catches have shown a sharp decline from the 20 million tonnes that was regularly landed in the 1980s, with the most recent harvest data (2004—12.6 million tonnes) being on a par with the quantities harvested in the early 1960s. The major contributor to this decline has been the collapse of the USSR—with Soviet landings dropping from 11 million tonnes in 1987 to just 2.5 million tonnes in 2004.

In Western Europe, the intensive exploitation of many demersal groundfish stocks (cod and haddock, for instance) during recent decades has meant that some stocks are now considered to be outside the safe biological limits and, as a consequence, European Union (EU) fleets are increasingly seeking access to other countries' exclusive economic zones (EEZs) for their distant-water fleets. The major (over 0.5 million tonnes) European marine fishing nations in 2004 were the Russian Federation, Norway, Iceland, Denmark, Spain, the Faeroe Islands and the United Kingdom.

A similar trend is evident in inland waters. Here catches have fallen from a late 1980s peak of 450 000 tonnes to 206 062 tonnes in 2004, the principal factor in the decline being a sharp fall in reported landings from the ex-Soviet economies.

The one success story has been aquaculture. In 2004, 2.24 million tonnes were produced at a value exceeding US$5.58 billion—up over 50% in value and volume terms since 1994—with farmed diadromous species (principally salmon and trout) accounting for the bulk of the growth and the major portion of current worth (57.5% of 2004 revenues). The main producing countries were Norway and, to a lesser extent, Scotland.

While current fish per caput supply levels in the region as a whole (at around 17 kg) are slightly above the global average, levels vary considerably among sub-regions and countries. While in the industrialized countries the average per caput supply is about 22.5 kg per year, in Iceland mean fish supply exceeds 90 kg per capita. In contrast, in Bulgaria, Rumania and Serbia-Montenegro, annual supply is less than 4 kg per capita.

Latin America and the Caribbean. In Latin America and the Caribbean, total marine landings are heavily influenced by fluctuations in small pelagic (principally anchoveta) stocks, and periodic stock collapses (such as the anchoveta collapses in 1973 and, to a lesser extent, in the early 1980s—and subsequently in 1998) result in sharply reduced harvests.

This has been slightly offset by an increase in demersal production (particularly hake) and squid in the southwest Atlantic and tuna across the region, but most major demersal and small pelagic stocks are now considered to be fully-or over-exploited. Marine production in the region peaked in 1994 when 22.8 million tonnes was landed, with current (2004) landings around 17.5 million tonnes. Major Latin American marine fishing nations (over 0.5 million tonnes) in 2004 were Peru, Chile, Mexico and Argentina.

In comparison, Latin American inland fisheries are rather less significant, yielding around 325 000 to 375 000 tonnes per annum over the last few decades, with the recent rise in recorded landings (to around 440 000 tonnes) more due to better recording of catches than the exploitation of new fisheries. The same cannot be said of aquaculture, however. The abundance of small pelagic fish provided the stimulus for the development of an important fish reduction industry (currently consuming more than two-thirds of the total marine catch) and with it, a domestic aquaculture industry. The contribution of aquaculture—in volume and value terms—increased by more than threefold over the decade to 2004, production increasing from 0.34 to 1.14 million tonnes while the value of aquacultural output grew from US$1.32 to 4.56 billion. The main impulse for this growth came from the expansion of freshwater fish and marine diadromous (essentially salmon and trout) farming—up five and six-fold respectively, with Chile exclusively responsible for the upsurge in diadromous output and Brazil largely accounting for the increase in freshwater fish landings. As in other regions, fish supply and consumption levels vary widely. In some of the small Caribbean island states (Grenada, Montserrat, the Turks and Caicos islands) per caput supply exceeds 40 kg p.a. —substantially higher than in either Central America (around 9.3 kg p.a.) or South America (8.8 kg p.a.)—while its contribution to daily animal protein requirements ranges from 1.7% in Bolivia to 60.5% in Belize.

Africa

African marine catches, having risen from 3 million tonnes in 1990, stabilised at around 4.8 million tonnes at the start of the twenty-first century. However, these aggregate figures mask a number of differing trends. Landings in North Africa, which accounted for around 26% of the 2004 catch, are dominated by Morocco (67% of sub-regional landings), which realizes 95% of its catch in the Atlantic. Contemporary evidence suggests there is little scope for further exploitation of Mediterranean fish stocks. Southern Africa harvests have grown above trend, enabling the region to increase its share of continental landings from 30 to 34% over the period 1990-2004.

Production is dominated by South Africa (52% of sub-regional landings), with its focus on the Cape hake fisheries, and Namibia (34%), where the emphasis is on small pelagic stocks. Eastern sea-board landings are much less important, accounting for under 4% of continental landings in 2004, a percentage that has shrunk moreover as landings in the sub-region have risen at well below trend rates since 1990. The Western sea-board supplied 36% of the marine catch in 2004, largely drawn from Senegal (374 245 tonnes in 2004), Ghana (313 935 tonnes) and Nigeria (251 232 tonnes). Although the Ghanaian catch in 2004 is of a similar magnitude to that recorded in 1990, Senegalese pelagic landings have increased dramatically (100 000+ tonnes) over the same time-span.

Africa has the world's most important inland fisheries, 2.25 million tonnes being harvested in 2004 compared to 1.8 million tonnes a decade previously. Here too though the aggregate figure obscures regional variations. Tanzania and Kenya, who source a large portion of their inland catch from the Lake Victoria Nile perch fishery have experienced landing declines of 16 and 38% respectively since 1990 (although Ugandan catches have remained rather more buoyant). Elsewhere, however, the story has been one of increased exploitation of inland water resources—catch estimates for the Democratic Republic of Congo suggest landings now stand at 217 000 tonnes (up 35.6% since 1990), Egypt has seen output leap by 52% (to 238,455 tonnes), while Nigerian landings have doubled to 215 000 tonnes.

Aquaculture, by contrast, is less well-developed in the region, output reaching 570 113 tonnes, worth US$893 million in 2004. Nevertheless, this represents a substantive improvement when set beside the equivalent 1990 figures of 82 475 tonnes and US$166 million. Egypt is the principal aquaculture producer, presently accounting for around 83% of volume and 69% of value, with common carp culture well to the fore. Other countries with briskly growing aquaculture sectors include Kenya, Madagascar, Nigeria, South Africa and Zambia, all of whom have doubled their annual production (admittedly, from a small initial base) several times over the last decade or so.

Per caput fish supply in the region is low by international standards. Sub-regional averages lie between 3.7 kg (East Africa) and 11.6 kg (West Africa) p.a., with Niger (1.6 kg) and Gambia (29.3 kg) representing the extremes. Fish, historically, has nonetheless been a popular food item, providing more than half an individual's daily animal protein needs in Gambia, Ghana and Sierra Leone and—since most parts of the fish are consumed—it has also contributed significantly to calcium and iodine intakes.

However, the growth of the fish trade with the EU (facilitated, in part, by fishing agreements between the EU and various African nation states), Japan and, more recently, China, is leading to a reduction in the domestic availability of fish in both the main West African nations and those landlocked Central African countries that used to import fish from West Africa.

North America

In North America, marine catches declined steadily from a peak of 6.8 million tonnes in 1988 to 5.6 million tonnes in 2004, principally due to the overexploitation of the main commercial groundfish stocks in some areas where fisheries are now closed or subject to restrictions.

The most notable closure was in 1992 when Canada closed its northern cod fisheries off the Grand Banks, and over 20,000 people lost their jobs. Canadian landings, as a consequence, have fallen rather more sharply (down just over 50% since 1990, to 614 107 tonnes) than US landings over the last decade or so.

A similar trend is evident in inland waters, the take dropping from a 1989 peak of 205 503 tonnes to the current level of 143 370 tonnes (down 30%), despite the growing popularity of sport fishing, most notably in USA where federal agencies have been particularly active in developing recreational fisheries programmes since 1995. In the aquaculture sector, however, production in both volume and value terms has more than doubled since the early 1990s (currently output is 955 178 tonnes worth US$1.99 billion).

Aquaculture in North America—unlike in Europe or South America—is a much more diversified industry which includes marine, freshwater and diadromous fishes, crustaceans, molluscs and plants, although salmon and trout dominate in Canada, while catfish is an important sub-component of the US industry. Per caput fish supply levels in North America—at 22.7 kg per head—are comparable to the levels recorded in the industrialized countries of Europe, and sharply higher than the rates posted in South and Central America.

Oceania

Oceania provides comparatively little of the global marine catch (<2%), landing 1.4 million tonnes in 2004 although this is certainly an underestimate as much of the coastal catch in the region's small island developing states (SIDS) goes unreported. Although catches have risen by around 53% over the past decade, concerns over the current status of major commercial stocks has prompted the introduction of individual transferable quotas (ITQs) in New Zealand, the main regional fishery. Besides New Zealand (44% of regional landings—cod and molluscs each accounting for a third of the NZ catch), the other major regional fishing nations are Australia (18%, a quarter of which comes from the crustacean catch) and Papua New Guinea (15%, almost exclusively tuna). However, for many SIDS it is a key strategic sector in the quest for development given its contribution to production, local food security, employment and export earnings.

Landings from inland waters are equally low in global terms, with a reported catch of just 11,313 tonnes in 2004—largely from Papua New Guinea (96% of total), where inland fisheries in the highland areas are an important source of food security. However, regional data does not presently include sport and recreational fisheries—important fisheries in both New Zealand and Australia. Oceania's contribution to global aquaculture output is also relatively small, amounting to 139,273 tonnes worth US$447 million in 2004, although it has grown by 86% and 166% in volume and value terms respectively over the past decade. Production is almost exclusively concentrated in New Zealand (66% of volume, 37% of value) and Australia (28% of volume, 59% of value) and, unlike in other regions, it is mollusc cultivation (principally mussels in New Zealand and oysters in Australia) rather than diadromous fish farming that dominates.

Fish are an important protein source for many Pacific islanders—supplying as much as 77% of the Solomon Islands' and 85% of the Maldives' animal protein needs—although culturally and nutritionally it is of lesser importance in both New Zealand and Australia. While under-reporting tends to bias supply figures downwards for many of the Oceanic countries, nevertheless 2004 per caput supply ascended to 185.9 kg in the Maldives (the highest in the world), 75.2 kg in Kiribati and 94.3 kg in Palau.

Index

A

Agro Food Processing sector, 28.
Animal Biotechnology, 215, 217, 222, 223.
Aquaculture Technology, 98.
Aquatic Plants, 8, 11, 62, 76, 81, 85, 104, 107, 108, 115, 120, 134, 202, 205, 253, 260.
Aquatic Products, 27, 28, 31.
Aquatic Resources, 61, 62, 196.
Aquatic Species, 197, 198, 199.
Artificial Propagation, 83, 86, 90.

B

Biotechnology Research, 134, 135, 145, 146, 147, 148, 149, 152, 219, 228, 229, 235, 236.

C

Cell Culture, 145, 201, 203, 205, 206, 207.
Chilled Fish, 34, 36, 39, 185.
Circulatory System, 37.
Cold Storage, 157, 159, 160, 161, 166.
Crop Cultivation, 99, 100, 101.
Crop Farming, 99.
Crop Integration, 99, 100, 101, 102, 103, 105, 106, 126, 127, 129.
Crop Variety, 101.
Cultivation Systems, 101.

D

Digestive Tract, 45.
DNA Technology, 132, 149, 151, 222, 224, 225, 226, 230.

E

Endocrine System, 90.
Energy Metabolism, 58.
Evaluate Production Efficiency, 78.

F

Fatty Acids, 44, 45, 49, 50, 55, 70, 208, 256.
Feed Production, 220.
Feeding Habits, 75, 76, 77, 82.
Fish Byproducts, 166.
Fish Catches, 27.
Fish Culture, 94, 95, 96, 98, 99, 109, 110, 112, 119, 120, 121, 128, 170, 241.
Fish Farming, 71, 94, 95, 96, 97, 98, 99, 100, 101, 102, 104, 105, 106, 109, 111, 112, 113, 114, 116, 117, 118, 119, 120, 121, 122, 126, 127, 128, 129, 130, 131, 262.
Fish Feeds, 95, 99, 101, 102, 104, 109, 118, 121, 127.
Fish Handling, 165, 167, 169, 171, 182, 183.
Fish Lipids, 44, 49.
Fish Muscle Tissue, 50, 56.
Fish Muscles, 187.
Fish Oils, 50, 166.
Fish Ponds, 71, 95, 97, 98, 99, 100, 103, 105, 106, 107, 108, 109, 110, 111, 112, 113, 114, 115, 116, 117, 119, 120, 121, 122, 124, 125, 126, 127, 129, 130, 241.

Fish Production, 63, 71, 96, 97, 150, 200, 203, 204, 205, 206, 243, 244, 245, 255.
Fish Products, 24, 28, 29, 30, 31, 32, 41, 69, 165, 167, 169, 172, 180, 195, 196, 197, 198.
Fish Quality, 183, 187.
Fish Species, 2, 4, 7, 11, 34, 36, 39, 42, 44, 47, 48, 49, 51, 52, 53, 59, 60, 61, 63, 64, 65, 66, 68, 115, 136, 167, 171, 186, 187, 188, 189, 195, 204, 219, 220, 223, 224, 225, 226, 234, 235.
Fish Yields, 99, 109, 110, 111, 116, 125.
Fisheries Development, 141.
Fisheries Sector, 27, 138, 142.
Fisheries Technology, 209.
Fishery Management, 195, 196, 197.
Fishery Resources, 30, 72, 196, 254.
Fishing Methods, 255.
Fishing Operations, 255.
Fishing Vessels, 12, 13, 16, 21, 24, 25, 157, 158, 159, 163, 168, 252.
Food Consumption, 27.
Food Industry, 179, 226.
Food Security, 150, 257, 262.
Forest Resources, 213.
Fresh Fish Handling, 169.
Frozen Storage, 43.

G

Genetic Resources, 132, 137, 139, 140, 141, 150, 195, 218, 227, 235, 236, 237, 238, 239.
Gonad Development, 89, 90, 91, 92.

H

Handling Shellfish, 165.

I

Integrated Farming, 71.
Integrated Fish Farming, 71, 94, 95, 96, 97, 98, 99, 109, 112, 116, 117, 119, 130, 131.
Integrated Fisheries Project, 31.
Intellectual Property Rights, 132, 133, 152, 219, 228, 232, 237.

M

Marine Biotechnology, 204, 205, 206, 210, 211.
Marine Fisheries, 25, 257.
Market Regime, 241, 242.
Muscle Proteins, 38, 43, 44, 51, 187.

N

Nervous System, 90, 91, 92.

O

Onboard Handling, 184.

P

Packing Materials, 167.
Plant Biotechnology, 212, 227, 228, 229, 236, 237.
Plant Genetic Resources, 238, 239, 241.
Proteolytic Enzymes, 40, 41.
Pseudomonas Spoilage, 69, 192.

Q

Quality Assessment, 34.

S

Seafood Processing, 32.
Seaweed Cultivation, 209.
Sensory Evaluation, 34, 36.
Specific Spoilage Organisms, 65, 155.
Spoilage Organisms, 65, 155.
Sustainable Fisheries, 13.
Sustainable Production, 140.
Swim Bladder, 51.

T

Tropical Fish, 33, 46, 64, 66, 67, 153, 155, 156, 191, 192, 193.

V

Vaccines Development, 206.